Lecture Notes in Computer Science 14547

The series Lecture Notes in Computer Science (LNCS), including its subseries Lecture Notes in Artificial Intelligence (LNAI) and Lecture Notes in Bioinformatics (LNBI), has established itself as a medium for the publication of new developments in computer science and information technology research, teaching, and education.

LNCS enjoys close cooperation with the computer science R & D community, the series counts many renowned academics among its volume editors and paper authors, and collaborates with prestigious societies. Its mission is to serve this international community by providing an invaluable service, mainly focused on the publication of conference and workshop proceedings and postproceedings. LNCS commenced publication in 1973.

Maria De Marsico · Gabriella Sanniti Di Baja · Ana Fred
Editors

Pattern Recognition Applications and Methods

12th International Conference, ICPRAM 2023 Lisbon, Portugal, February 22–24, 2023 Revised Selected Papers

Editors
Maria De Marsico
Sapienza University of Rome
Rome, Italy

Gabriella Sanniti Di Baja
ICAR-CNR
Napoli, Italy

Ana Fred
Instituto de Telecomunicações
Lisbon, Portugal

University of Lisbon
Lisbon, Portugal

ISSN 0302-9743 ISSN 1611-3349 (electronic)
Lecture Notes in Computer Science
ISBN 978-3-031-54725-6 ISBN 978-3-031-54726-3 (eBook)
https://doi.org/10.1007/978-3-031-54726-3

This Springer imprint is published by the registered company Springer Nature Switzerland AG
The registered company address is: Gewerbestrasse 11, 6330 Cham, Switzerland

Paper in this product is recyclable.

Preface

The present book includes extended and revised versions of a set of selected papers from the 12th International Conference on Pattern Recognition Applications and Methods (ICPRAM 2023), held in Lisbon, Portugal, from 22 to 24 February 2023.

The International Conference on Pattern Recognition Applications and Methods is a major point of contact between researchers, engineers and practitioners in the areas of Pattern Recognition and Machine Learning, both from theoretical and application perspectives. Contributions describing applications of Pattern Recognition techniques to real-world problems, interdisciplinary research, and experimental and/or theoretical studies yielding new insights that advance Pattern Recognition methods are especially encouraged.

ICPRAM 2023 received 157 paper submissions from 40 countries, of which 8 papers were included in this book.

The papers were selected by the event chairs and their selection was based on a number of criteria that included the classifications and comments provided by the program committee members, the session chairs' assessment and also the program chairs' global view of all papers included in the technical program. The authors of selected papers were then invited to submit revised and extended versions of their papers having at least 30% innovative material.

The papers selected to be included in this book contribute to the understanding of relevant trends of current research on Pattern Recognition Applications and Methods, including: Classification and Clustering, Biometrics, Image and Video Analysis and Understanding, Knowledge Acquisition and Representation, Deep Learning and Neural Networks, Information Retrieval, Industry-Related Applications, Feature Selection and Extraction, Medical Imaging and Document Analysis.

We would like to thank all the authors for their contributions and also the reviewers who have helped to ensure the quality of this publication.

February 2023

Maria De Marsico
Gabriella Sanniti Di Baja
Ana Fred

Organization

Conference Chair

Ana Fred	Instituto de Telecomunicações and University of Lisbon, Portugal

Program Co-chairs

Maria De Marsico	Sapienza Università di Roma, Italy
Gabriella Sanniti Di Baja	Italian National Research Council CNR, Italy

Program Committee

Andrea Abate	University of Salerno, Italy
Rahib Abiyev	Near East University, Turkey
Paolo Addesso	Università degli Studi di Salerno, Italy
Heba Afify	Cairo University, Egypt
Mayer Aladjem	Ben-Gurion University of the Negev, Israel
José Alba	University of Vigo, Spain
George Azzopardi	University of Groningen, The Netherlands and University of Malta, Malta
Antonio Bandera	University of Málaga, Spain
Lluís A. Belanche	BarcelonaTech, Spain
Stefano Berretti	University of Florence, Italy
Monica Bianchini	University of Siena, Italy
Ronald Böck	Genie Enterprise Inc., Germany
Andrea Bottino	Politecnico Di Torino, Italy
Paula Brito	Universidade do Porto, Portugal
Alfred Bruckstein	Technion, Israel
Javier Calpe	Universitat de València, Spain
Modesto Castrillon-Santana	Universidad de Las Palmas de Gran Canaria, Spain
Amitava Chatterjee	Jadavpur University, India
Gerard Chollet	CNRS, France
Luiza de Macedo Mourelle	State University of Rio de Janeiro, Brazil
José Joaquim de Moura Ramos	University of A Coruña, Spain

Yago Diez	Yamagata University, Japan
Jean-Louis Dillenseger	Université de Rennes, France
Mounin El Yacoubi	Institut Polytechnique de Paris, France
Giorgio Fumera	University of Cagliari, Italy
Markus Goldstein	Ulm University of Applied Sciences, Germany
Petra Gomez-Krämer	La Rochelle University, France
Bernard Gosselin	University of Mons, Belgium
Michal Haindl	Institute of Information Theory and Automation, Czech Republic
Kouichi Hirata	Kyushu Institute of Technology, Japan
Sean Holden	University of Cambridge, UK
Paul Honeine	University of Rouen Normandy, France
Su-Yun Huang	Academia Sinica, Taiwan, Republic of China
Dimitris Iakovidis	University of Thessaly, Greece
Akinori Ito	Tohoku University, Japan
Yuji Iwahori	Chubu University, Japan
Sarangapani Jagannathan	Missouri University of Science and Technology, USA
Mohammad R. Jahanshahi	Purdue University, USA
Arti Jain	Jaypee Institute of Information Technology, India
Roger King	Mississippi State University, USA
Pavel Kordík	Czech Technical University in Prague, Czech Republic
Marcin Korzen	West Pomeranian University of Technology in Szczecin, Poland
Sotiris Kotsiantis	University of Patras, Greece
Kidiyo Kpalma	Institut National des Sciences Appliquées de Rennes, France
Marek Kretowski	Bialystok University of Technology, Poland
Adam Krzyzak	Concordia University, Canada
Young-Koo Lee	Kyung Hee University, South Korea
Josep Llados	Universitat Autònoma de Barcelona, Spain
Eduardo Lleida	Universidad de Zaragoza, Spain
Javier Lorenzo-Navarro	Universidad de Las Palmas de Gran Canaria, Spain
Pasi Luukka	Lappeenranta University of Technology, Finland
Delia Mitrea	Technical University of Cluj-Napoca, Romania
Valeri Mladenov	Technical University of Sofia, Bulgaria
Ramón Mollineda Cárdenas	Universitat Jaume I, Spain
Muhammad Marwan Muhammad Fuad	Coventry University, UK
Marco Muselli	Consiglio Nazionale delle Ricerche, Italy
Jakub Nalepa	Silesian University of Technology, Poland

Mita Nasipuri	Jadavpur University, India
Mikael Nilsson	Lund University, Sweden
Mark Nixon	University of Southampton, UK
Kalman Palagyi	University of Szeged, Hungary
Shahram Payandeh	Simon Fraser University, Canada
Mikhail Petrovskiy	Lomonosov Moscow State University, Russian Federation
Vincenzo Piuri	Università degli Studi di Milano, Italy
Horia Pop	Babes-Bolyai University, Romania
Sivaramakrishnan Rajaraman	National Library of Medicine, USA
Daniel Riccio	University of Naples, Federico II, Italy
Luciano Sanchez	Universidad de Oviedo, Spain
Antonio-José Sánchez-Salmerón	Universitat Politècnica de València, Spain
Carlo Sansone	University of Naples Federico II, Italy
Michele Scarpiniti	Sapienza University of Rome, Italy
Gerald Schaefer	Loughborough University, UK
Mu-Chun Su	National Central University, Taiwan, Republic of China
Eulalia Szmidt	Systems Research Institute, Polish Academy of Sciences, Poland
Monique Thonnat	Inria, France
Kar-Ann Toh	Yonsei University, South Korea
Edmondo Trentin	Università degli Studi di Siena, Italy
Pei-Wei Tsai	Swinburne University of Technology, Australia
Ernest Valveny	Universitat Autònoma de Barcelona, Spain
Asmir Vodencarevic	Daiichi Sankyo Europe GmbH, Germany
Laurent Wendling	Université Paris Cité, France
Slawomir Wierzchon	Polish Academy of Sciences, Poland
Michal Wozniak	Wroclaw University of Science and Technology, Poland
Jing-Hao Xue	University College London, UK
Pavel Zemcik	Brno University of Technology, Czech Republic
Leishi Zhang	Canterbury Christ Church University, UK

Additional Reviewers

Paolo Andreini	University of Siena, Italy
Jeongik Cho	Concordia University, Canada
Filippo Costanti	University of Florence, Italy
Rita Delussu	University of Cagliari, Italy
Veronica Lachi	University of Siena, Italy

Hai Wang	University College London, UK
Xiaoke Wang	University College London, UK

Invited Speakers

Petia Radeva	Universitat de Barcelona, Spain
Anil Jain	Michigan State University, USA
Mario Figueiredo	Instituto de Telecomunicações, Portugal

Contents

Theory and Methods

Exploring Data Augmentation Strategies for Diagonal Earlobe Crease Detection

Sara Almonacid-Uribe, Oliverio J. Santana, Daniel Hernández-Sosa, and David Freire-Obregón(✉)

SIANI, Universidad de Las Palmas de Gran Canaria, Las Palmas de Gran Canaria, Spain
david.freire@ulpgc.es

Abstract. Diagonal earlobe crease (DELC), also known as Frank's sign, is a diagonal crease, line, or deep fold that appears on the earlobe and has been hypothesized to be a potential predictor of heart attacks. The presence of DELC has been linked to cardiovascular disease, atherosclerosis, and increased risk of coronary artery disease. Some researchers believe that DELC may be an indicator of an impaired blood supply to the earlobe, which could reflect similar issues in the heart's blood supply. However, more research is needed to determine whether DELC is a reliable marker for identifying individuals at risk of heart attacks or other cardiovascular problems. To this end, the authors have released the first DELC dataset to the public and investigated the performance of numerous state-of-the-art backbones on annotated photos. The experiments demonstrated that combining pre-trained encoders with a customized classifier achieved 97.7% accuracy, with MobileNet being the most promising encoder in terms of the performance-to-size trade-off.

Keywords: Computer vision · Diagonal earlobe crease · DELC · Frank's sign · Cardiovascular disease · Coronary artery disease · deep learning

1 Introduction

Heart disease is a major public health concern in the world. The Centers for Disease Control and Prevention (CDC) reports that it is the leading cause of death for both men and women, as well as the majority of ethnic and racial groups in the country [6]. Shockingly, cardiovascular disease is responsible for one death every 34 s in the United States, making it a significant contributor to mortality rates [6]. Moreover, it's worth noting that one in five heart attacks is silent, with the individual unaware of the damage done [30]. Therefore, early detection is crucial for managing symptoms, reducing mortality rates, and improving the quality of life [4]. This highlights the need for preventative measures, including lifestyle changes and regular checkups, to identify and address potential risk factors before they lead to more severe health issues (Fig. 1).

We want to acknowledge Dr. Cecilia Meiler-Rodríguez for her creative suggestions and inspiring ideas. This work is partially funded by the ULPGC under project ULPGC2018-08, by the Spanish Ministry of Science and Innovation under project PID2021-122402OB-C22, and by the ACIISI-Gobierno de Canarias and European FEDER funds under projects ProID2020010024, ProID2021010012, ULPGC Facilities Net, and Grant EIS 2021 04.

M. De Marsico et al. (Eds.): ICPRAM 2023, LNCS 14547, pp. 3–18, 2024.
https://doi.org/10.1007/978-3-031-54726-3_1

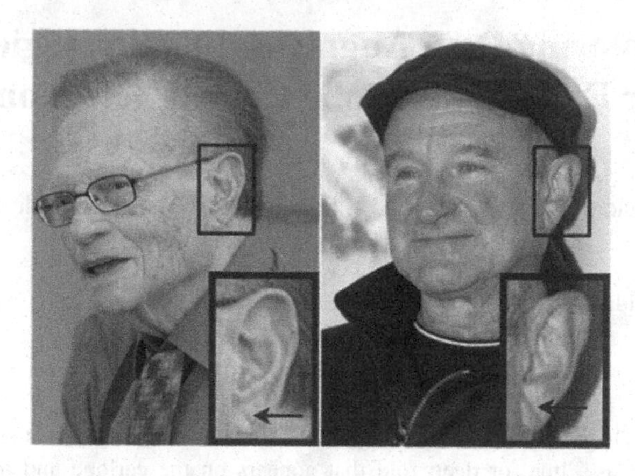

Fig. 1. Celebrities exhibiting a DELC marker. In 1987, the former CNN interviewer Larry King experienced a heart attack and subsequently underwent bypass surgery (photo credit: Eva Rinaldi, Wikimedia Commons, CC-BY-SA 2.0). In 2009, Robin Williams, the former comedian and actor, underwent aortic valve replacement surgery (photo credit: Angela George, Wikimedia Commons, CC-BY 3.0). In both pictures, the ear is prominently highlighted [1].

Clinicians typically diagnose coronary artery disease (CAD) based on medical history, biomarkers, raw scores, and physical examinations. However, with technological advancements, the diagnostic process has evolved. Deep learning (DL) has shown tremendous potential in detecting abnormalities in computed tomography (CT) images over the last decade [3]. Several DL techniques have been proposed to automatically estimate CAD markers from CT images. These models predict clinically relevant image features from cardiac CT, such as coronary artery calcification scoring [18,33,35], localization of non-calcified atherosclerotic plaque [34,37], and stenosis from cardiac CT [20,38]. The development of these DL models has the potential to revolutionize the way CAD is diagnosed and treated by providing more accurate and efficient diagnostic tools. Clinicians can use these models to identify early signs of CAD, enabling timely interventions that can help prevent the progression of the disease and improve patient outcomes.

The use of DL in detecting abnormalities in CT images has shown great promise in diagnosing cardiac diseases. However, it comes with a steep cost due to the expensive CT equipment. Additionally, cardiac illnesses are often difficult to detect unless patients exhibit symptoms and seek medical attention for cardiac checkups. This is where DELC is a potentially helpful tool for identifying cardiac problems.

DELC is a visible crease that runs diagonally from the tragus to the border of the earlobe and it is commonly associated with atherosclerosis, particularly CAD. Frank first described it in a case series of CAD patients; hence it is also known as Frank's Sign [11]. Since then, numerous studies have been published on its potential use as a diagnostic tool for identifying cardiac diseases [32]. Despite being less well-known than traditional approaches, DELC examinations are painless, non-invasive, and easy to interpret. If the diagnostic accuracy of DELC examinations is proven sufficient for decision-making, they could be used in primary care or emergency departments as a

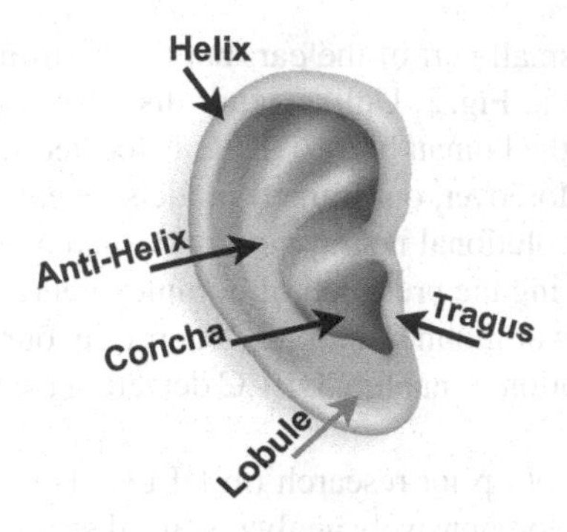

Fig. 2. Outer ear scheme. The earlobe, also known as the lobulus auriculae, comprises areolar and adipose connective tissues, giving it a soft and flexible structure. Unlike the rest of the ear, which contains cartilage, the earlobe lacks rigidity. This unique composition allows for a rich blood supply, contributing to its role in regulating the temperature of the ears and aiding in warmth retention [27].

cost-effective and accessible method for identifying individuals at risk for cardiac diseases. To explore the potential of DELC as a diagnostic tool, we have developed a DELC detector using state-of-the-art DL backbones and ear collections as benchmarks for the models. Firstly, we gathered DELC ear images available on the Internet. Then we developed multiple DL models that considered pre-trained encoders or backbones to predict whether an ear displays DELC. Moreover, we analyzed the performance of the different backbones by varying the classifier parameters and found no correlation between the number of parameters and the model quality. By developing a DL-based DELC detector and exploring the diagnostic accuracy of DELC examinations, we hope to provide healthcare professionals with a valuable tool for identifying individuals at risk for cardiac diseases, particularly those who may not have access to expensive CT equipment or exhibit symptoms of heart disease.

In addition, we conducted an extensive evaluation of the considered backbones by experimenting with different classifier parameters and employing two distinct augmentation strategies. Interestingly, our findings revealed a lack of correlation between the number of parameters and the performance of the best-performing model. This observation emphasizes the intricate nature of the DELC detection problem and underscores the significance of adopting comprehensive evaluation methods.

We evaluated our proposal using a dataset that combined images from different sources. The dataset comprises 342 positive DELC images by gathering publicly available images from the Internet and cropping out the ears. For negative samples, the dataset provides ear images of another publicly accessible ear database called AWE [10]. The images were collected from various natural settings, including lighting, pose, and shape variations. Given the relatively small number of samples, we employed data augmentation techniques during the training process. Our results were remarkable, with predictions being accurate up to 97.7%, and we gained some interesting insights from our findings.

The earlobe may be a small part of the ear, but the performance of our classifier is noteworthy, as highlighted in Fig. 2. Unlike other diseases, such as melanoma, which can be found anywhere in the human body, DELC is located in a specific area, making the detection task easier. Moreover, our trained models reveal an interesting behaviour. By using light-weight convolutional neural networks such as MobileNet, we achieved high accuracy while balancing the precision of complex neural network structures with the performance constraints of mobile runtimes. Therefore, our proposal could be beneficial for ubiquitous applications, enabling DELC detection using a smartphone anytime and anywhere.

This paper builds upon our prior research on DELC [1] and introduces several key contributions. Firstly, it comprehensively analyzes two distinct augmentation strategies, namely soft and hard, as practical solutions for addressing the DELC detection challenge. This analysis provides valuable insights into each strategy's potential advantages and trade-offs. Secondly, we demonstrate the effectiveness of combining pre-trained backbones with a novel classifier to achieve remarkable accuracy in DELC prediction. By leveraging the knowledge encoded in the pre-trained backbones, we enhance the model's ability to detect and classify DELC patterns accurately. Furthermore, our findings align with a size-performance trade-off analysis, demonstrating that even lightweight encoders can be employed without sacrificing accuracy in DELC detection. This significant discovery presents exciting prospects for the widespread adoption of this technology, as it enables the development of efficient and precise DELC detection systems.

The rest of this paper is structured as follows. In the following section, we discuss previous related work. The used dataset is described in Sect. 3. In Sect. 4, we describe the proposed pipeline, which details our methodology for DELC detection. In Sects. 5 and 6, we report on the experimental setup and results, including a detailed analysis of the performance of the different backbones. Finally, in Sect. 7, we draw conclusions and discuss future research directions.

2 Related Work

An in-depth understanding of the current state of the art in the field of DELC detection requires an examination from both physiological and technological perspectives. Physiological studies investigate the relationship between CAD and DELC and provide evidence to support this association. Meanwhile, technological research evaluates proposals from the Computer Vision community for detecting DELC using machine learning algorithms. By combining these two perspectives, researchers can comprehensively understand the current solutions to the problem and identify opportunities for future advancements.

2.1 Physiological Perspectives on DELC

DELC is characterized by a diagonal fold in the earlobe that forms at a 45° angle from the intertragic notch to the posterior edge of the ear. Multiple studies have found a correlation between DELC and cardiac problems in recent decades. A grading system has been established based on various features, such as length, depth, bilateralism, and inclination, which are linked to the incidence of CAD. Complete bilateralism is considered the most severe form of DELC [23].

Frank first proposed the association between DELC and CAD [11]. While some scientists suggest that DELC indicates physiological aging [21], the CAD hypothesis has gained support from subsequent research. Studies have shown that the presence of DELC can accurately predict the likelihood of cardiovascular issues in patients. Given that coronary disease is a leading cause of death in developed countries [24], early detection is crucial in improving future patient quality of life and preventing or reducing CAD-related mortality.

In a pioneering study, Pasternac et al. evaluated 340 patients, of whom 257 had CAD [22]. The study revealed that 91% of patients with DELC had CAD, making it the most prevalent sign in those with more severe disease. More recently, Stoyanov et al. conducted a study involving 45 patients, 16 females, and 29 males [28]. Of these, 22 individuals had a well-documented clinical history of CAD, while the remaining patients did not. During general examination before the autopsy, 35 patients had well-formed DELC. Furthermore, patients with pierced ears did not show any signs of lobule injury due to piercing, and the observed creases were therefore accepted as DELC.

2.2 Computer Vision Perspectives on DELC

In Computer Vision, biometric traits have been extensively studied over the years. More recent research has delved into areas such as gait analysis and body components to tackle diverse tasks, including facial expressions [25] and face/voice verification [13]. Additionally, ear recognition has garnered significant attention [2, 12].

In healthcare applications, several notable proposals have emerged for diagnosing various ear-related diseases, including otitis media, attic retraction, atelectasis, tumors, and otitis externa. These conditions encompass most ear illnesses typically diagnosed by examining the eardrum through otoendoscopy [7]. Notably, Zeng et al. recently achieved a 95.59% accuracy in detecting some diseases by combining multiple pre-trained encoders and utilizing otoendoscopy images as input [36]. The authors argued that leveraging pre-trained deep learning architectures offers a significant advantage over traditional handcrafted methods. To diagnose chronic otitis, Wang et al. proposed a deep learning system that automatically extracted the region of interest from two-dimensional CT scans of the temporal bone [31]. Remarkably, the authors claimed that their model's performance (83.3% sensitivity and 91.4% specificity) was comparable to, and in certain cases superior to, that of clinical specialists (81.1% sensitivity and 88.8% specificity).

In our research, we have also adopted a deep learning approach to address the DELC detection problem. However, we focus on utilizing the ear as a marker for Computer-Aided Diagnosis.

In a published experimental study by Hirano et al., the focus was on analyzing the detection of DELC using Computer-Aided Diagnosis [15]. The authors employed a handcrafted approach, which involved manually trimming earlobes and utilizing a Canny edge detector to identify DELC in carefully captured images. Their experiment

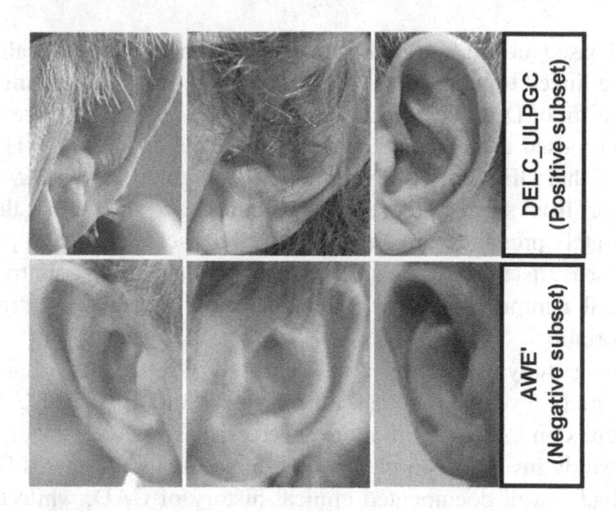

Fig. 3. DELC_ULPGC+AWE' Dataset. The studied dataset comprises two in-the-wild subsets. Both of them are gathered from the Internet: a subset of the well-known AWE dataset [10] as the DELC negative subset and the DELC_ULPGC subset [1].

involved photographing the ears of 88 participants from a single frontal angle, with only 16% of the participants being healthy.

In contrast to their study, our approach considered images of ears captured in real-world settings, which exhibited significant variations in pose and illumination. By addressing these challenges, we aimed to enhance the robustness and practicality of DELC detection in diverse environments.

3 Dataset

We have considered the DELC_ULPGC+AWE' Dataset, dedicated explicitly to DELC [1]. The collection of images provided by this dataset was acquired from the Internet under unrestricted conditions. As a result, these images exhibit significant pose, scale, and illumination variations, as depicted in Fig. 3. This dataset is mixed, since it comprises data from several sources. Concretely, it follows a four-step procedure, which includes the following:

- Web scraping. The authors conducted a targeted search using keywords such as "DELC", "Frank's Sign", and names of various celebrities to download relevant images from the web.
- Ear labeling. The labeled dataset involved trimming the ear region using the labelImg tool[1]. This labeling process resulted in a subset of 342 positive DELC images. In contrast to earlier handcrafted techniques [15], deep learning algorithms can detect patterns across larger regions. Consequently, the entire ear was considered for this dataset.

[1] https://github.com/heartexlabs/labelImg.

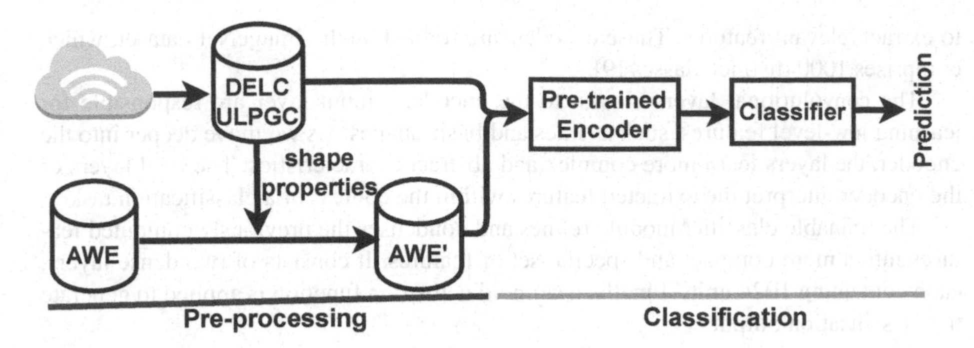

Fig. 4. The proposed pipeline for the DELC detection system. The proposed process consists of two primary modules: the ear pre-processing module and the classification module. The DELC_ULPGC+AWE' dataset is generated and prepared in the ear pre-processing module. This dataset is then passed to the classification module, where features are computed from the data. The resulting tensor serves as the input to the classifier, completing the overall process [1].

- Statistical shape analysis. Almonacid-Uribe et al. computed statistical shape properties, specifically the mean and standard deviation, from the positive subset of DELC images. This information was later utilized to obtain a negative DELC subset. The mean shape of the positive subset was determined to be 82×159 pixels.
- Generation of a negative subset. Finally, the authors utilized the publicly available AWE dataset [10] to create a negative subset. The AWE dataset includes "ear images captured in the wild" and Internet-based celebrity photos. Each subject in the dataset has ten photos, with sizes ranging from 15×29 pixels for the smallest sample to 473×1022 pixels for the largest. The AWE dataset was selected due to similarities related to the gathering process. Finally, images were extracted from the AWE dataset that matched the resolution of the DELC-positive subset. To ensure that no DELC-positive samples were included, Almonacid-Uribe et al. carefully examined the selected images, resulting in a subset of 350 negative images called the AWE' subset.

4 Proposal Description

This paper introduces and evaluates a sequential training pipeline encompassing two main modules: a dataset generation module and a module incorporating a pre-trained backbone and a trainable classifier. The classifier calculates a distance measure based on the generated embeddings, and this distance metric is utilized to compute a loss function, which is then fed back to the classifier for training. Figure 4 visually represents the described approach.

4.1 Proposed Architecture

The implemented encoding process involves transforming the input data into a feature vector. Initially, each input sample is passed through an encoder that has been trained to extract relevant features. These encoders are trained on the ImageNet dataset, which comprises 1000 distinct classes [9].

The convolutional layers closest to the encoder's input layer are responsible for learning low-level features, such as lines and basic shapes. As we move deeper into the encoder, the layers learn more complex and abstract characteristics. The final layers of the encoder interpret the extracted features within the context of a classification task.

The trainable classifier module refines and condenses the previously computed features into a more compact and specific set of features. It consists of two dense layers, each containing 1024 units. Finally, a sigmoid activation function is applied to generate the classification output.

4.2 Data Augmentation

The available dataset needs a sufficient number of samples, with around 350 samples per class, which is inadequate to train a classifier without risking overfitting. To address this limitation, a data augmentation strategy was employed, generating 2100 augmented photos per class. Various transformations were applied to augment the dataset [5]. These augmented subsets were exclusively utilized for training, providing a more diverse and extensive training dataset. In this regard, employing both hard and soft augmentation techniques is crucial, since they play a pivotal role in enhancing the performance and robustness of the model, specifically in the context of earlobe crease detection.

Hard augmentation techniques, such as RandomBrightness, RandomContrast, HorizontalFlip, ShiftScaleRotate, HueSaturationValue, RGBShift, and RandomBrightnessContrast transformations, involve applying various transformations to the input data using input parameters. In the case of the hard augmentation techniques, these parameters were higher. By utilizing input parameters, RandomBrightness and RandomContrast can adjust the brightness and contrast levels, generating diverse lighting conditions that emulate real-world scenarios. HorizontalFlip horizontally flips the images, simulating the possibility of encountering earlobe creases on either side of a person's face. ShiftScaleRotate applies translations, scaling, and rotations to facilitate the model in learning invariant representations of diagonal earlobe creases. HueSaturationValue and RGBShift modify the color attributes of the images, introducing further variability in the training data. RandomBrightnessContrast combines changes in brightness and contrast to simulate a broader range of lighting conditions. With higher input parameters than the one used for the soft augmentation techniques, these hard augmentation techniques expose the model to a broader array of variations, enabling it to generalize better and robustly detect diagonal earlobe creases across different individuals and circumstances.

On the other hand, soft augmentation techniques, including rotation, RandomBrightness, RandomContrast, HueSaturationValue, RGBShift, and RandomBrightnessContrast, involve generating additional synthetic data samples. These techniques allow for creating new examples with realistic variations that may not be present in the original dataset. Rotation, for instance, helps simulate variations in the orientation of diagonal earlobe creases, as different individuals may exhibit creases at different angles. RandomBrightness, RandomContrast, HueSaturationValue, RGBShift, and RandomBrightnessContrast can be applied to the synthetic samples, introducing lighting, color, and intensity variations. By incorporating these soft augmentation techniques, the model is exposed to a broader range of realistic scenarios, enabling it to learn the distinguishing features of diagonal earlobe creases and discriminate them from other features that may resemble them. Soft augmentation also contributes to mitigating overfitting and improving the model's generalization by expanding the size and diversity of the training data. Combining hard and soft augmentation techniques, we create a comprehensive and representative dataset for training and evaluating the deep learning architecture for diagonal earlobe crease detection. Hard augmentation techniques simulate various real-world variations, while soft augmentation techniques generate additional samples with specific variations. Together, these techniques ensure that the model learns to detect diagonal earlobe creases accurately and robustly, even in the presence of noise, artifacts, or imperfect data. Moreover, integrating hard and soft augmentation techniques helps improve the model's resilience, reliability, and generalization capability, making it more effective in real-world scenarios where diagonal earlobe creases vary in appearance, orientation, or intensity.

Therefore, combining hard and soft augmentation techniques is essential for achieving optimal performance in detecting diagonal earlobe creases using deep learning architectures.

4.3 Backbone Comparison

For comparison purposes, several popular backbone models were considered: VGGNet [26], InceptionV3 [29], ResNet [14], Xception [8], MobileNet [16], and DenseNet [17]. These models have demonstrated strong performance in various computer vision tasks and were initially trained on the ImageNet dataset, which comprises 1000 distinct classes [9]. Using pre-trained encoders trained on ImageNet provides a valuable starting point, as these models have learned rich representations that can be leveraged for the diagonal earlobe crease detection task.

In our pipeline, as depicted in Fig. 4, each pre-trained encoder was coupled with a trainable classifier. The encoder, also called the backbone, captures the underlying features from the input images, while the classifier makes the final prediction based on these extracted features. This separation allows us to leverage the pre-trained weights of the encoder while fine-tuning the classifier to adapt it specifically for diagonal earlobe crease detection.

The training process involved optimizing the model's performance using the Adam optimizer [19]. This optimization algorithm is convenient for our proposal because it adapts the learning rate for each parameter individually, allowing for faster convergence and improved training efficiency. A learning rate of 10^{-3} was chosen to balance the model's ability to learn from the training data without overshooting the optimal solution. Additionally, a decay rate of 0.4 was applied, gradually reducing the learning rate over time to fine-tune the model further.

By employing these training settings and optimization techniques, we aim to maximize the performance and generalization capability of the deep learning models for diagonal earlobe crease detection. The pre-trained encoders offer a strong foundation of learned features, while the trainable classifiers enable the models to adapt specifically to the target task. By incorporating these elements and leveraging the ImageNet dataset, a strong foundation is established for training models with the capability to accurately identify diagonal earlobe creases.

5 Experimental Setup

In this section, we present two subsections that cover the setup and results of our experiments. The first subsection provides detailed technical information about our proposal, including the loss function and data split. The second subsection summarizes the achieved results.

Our detection scenario considers two outcomes for the sample classification: DELC and not DELC. Since it is a binary classification problem, we have employed the binary cross-entropy loss function to address the task:

$$Loss = -1/N * \sum_{i=1}^{N} -(y_i \log(p_i) + (1 - y_i) \log(1 - p_i)) \quad (1)$$

where p_i represents the i-th scalar value in the model output, y_i denotes the corresponding target value, and N represents the total number of scalar values in the model output.

Regarding the results, this section presents the average accuracy based on five repetitions of 9-fold cross-validation. On average, 615 original samples are utilized for training, while the remaining 69 samples are reserved for testing. It is worth noting that the selected training samples undergo data augmentation during the training process, enhancing the diversity and variability of the training set.

Table 1. Absolute comparison of different backbones on the DELC_ULPGC+AWE' dataset. The table is organized in terms of backbone, validation accuracy (Val. Acc.) and test accuracy (Test Acc.) for each augmentation approach, and the number of parameters of the backbone ($\#B_{Param}$). The bold entries show the best result and the lightest model.

Backbone	Soft Augmentation		Hard Augmentation		$\#B_{Param}$
	Val. Acc.↑	Test Acc.↑	Val. Acc.↑	Test Acc.↑	
Xception	95, 1%	94, 1%	97, 7%	95, 8%	22.9M
VGG16	96, 5%	93, 9%	91, 9%	85, 9%	138.4M
VGG19	95, 1%	92, 7%	93, 5%	88, 3%	143.7M
ResNet50	98, 1%	95, 8%	88, 1%	84, 7%	25.6M
ResNet101	97, 5%	94, 8%	93, 7%	88, 2%	44.7M
ResNet152	97, 8%	95, 1%	93, 4%	85, 7%	60.4M
MobileNet	98, 7%	96, 7%	95, 7%	93, 7%	**4.3M**
InceptionV3	98, 9%	**97, 7%**	97, 1%	96, 3%	23.9M
DenseNet121	96, 4%	95, 5%	90, 9%	88, 4%	8.1M
DenseNet169	88, 7%	88, 1%	93, 1%	90, 9%	14.3M
DenseNet201	95, 1%	93, 4%	96, 7%	95, 1%	20.2M

6 Results

Table 1 shows that in evaluating the performance of deep learning models for diagonal earlobe crease detection, we explored both hard and soft augmentation approaches. After thorough experimentation and analysis, it was observed that the soft augmentation approach yielded slightly better results compared to the hard augmentation approach.

The soft augmentation approach proved advantageous in enhancing the model's performance. By introducing realistic variations that may not exist in the original dataset, the soft augmentation approach allowed the model to learn to discriminate between true diagonal earlobe creases and other features that may resemble them. This additional exposure to diverse variations enabled the model to generalize better and improve its accuracy in detecting diagonal earlobe creases.

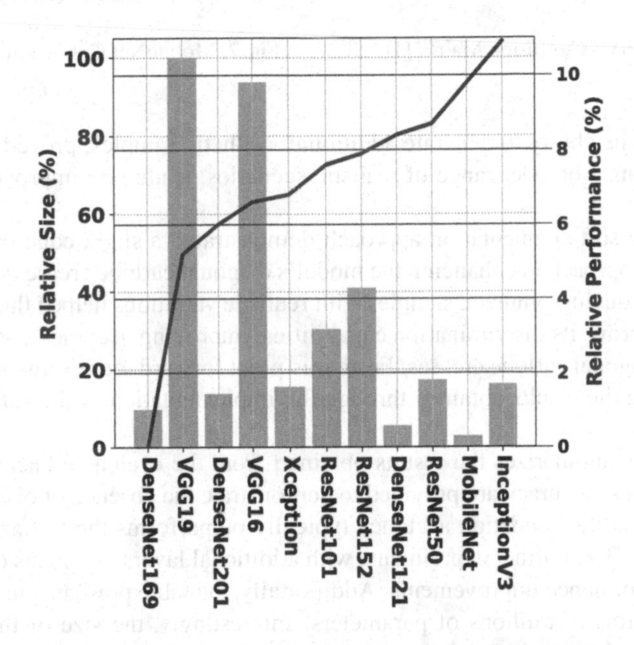

Fig. 5. Relative comparison performance and size of encoders considering the sof augmentation approach. The blue line in the graph represents the relative performance of the encoders, indicating higher values for better performance. On the other hand, the red bars represent the relative size of the encoders, with smaller bars indicating lighter models compared to those with higher bars [1] (Color figure online).

On the other hand, while the hard augmentation approach provided valuable variations to the training data, the results obtained were slightly inferior to the soft augmentation approach. The hard augmentation techniques primarily focused on modifying the existing images by altering brightness, contrast, flipping, scaling, and rotations. While these transformations introduced diversity, they may have yet to capture the range of variations encountered in real-world scenarios fully. As diagonal earlobe creases exhibit significant variations in appearance, orientation, and intensity, the soft augmentation

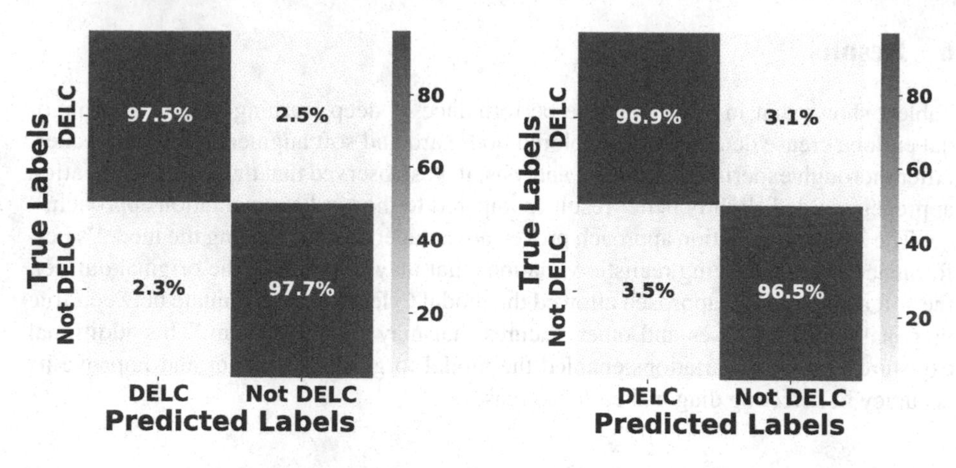

Fig. 6. InceptionV3 Confusion Matrix [1].

Fig. 7. MobileNet Confusion Matrix [1].

approach, with its ability to generate additional synthetic samples, proved more effective in simulating a broader range of realistic scenarios, leading to improved detection performance.

Overall, the soft augmentation approach demonstrated a slight edge over the hard augmentation approach in enhancing the model's diagonal earlobe crease detection performance. Introducing synthetic samples with realistic variations helped the model generalize and improve its discrimination capabilities, improving accuracy and robustness in detecting diagonal earlobe creases. From this point forward, our exclusive focus will be on analyzing the results obtained through the implementation of the soft augmentation approach.

Table 1 also summarizes the results obtained from the evaluated backbones. Both validation and test accuracy are provided to demonstrate the absence of overfitting. It is worth noting that the validation accuracy typically outperforms the test accuracy by a margin of 1% to 3%. Further experiments with additional layers or epochs did not yield significant performance improvements. Additionally, the table presents the size of each backbone in terms of millions of parameters. Interestingly, the size of the backbone does not directly impact the model's robustness.

Upon closer examination of the table, it becomes apparent that most pre-trained encoders exhibit impressive performance in detecting DELC. InceptionV3 achieves the highest accuracy of 97.7%, while DenseNet169 obtains the lowest accuracy of 88.1%.

Figure 5 comprehensively explores the relationship between different backbones by illustrating their relative performance and size. From a robustness perspective, it is evident that InceptionV3 outperforms DenseNet169 by more than 11%, making it the superior backbone. VGG19 and VGG16 are the largest models, while the other backbones are approximately 20% to 40% smaller.

Furthermore, Fig. 5 highlights the relative importance of size when addressing the DELC problem. It is observed that within each family of encoders, the smaller model tends to achieve better performance. For example, DenseNet121 outperforms other DenseNet models, VGG16 outperforms VGG19, and ResNet50 outperforms ResNet152 and ResNet101.

The insights from the top two models further support this observation. Despite a mere 1% difference in accuracy, the MobileNet encoder is significantly smaller (five and a half times) than the InceptionV3 encoder. Consequently, MobileNet performs three times faster than InceptionV3. However, a more detailed error analysis is required to assess the models' performance regarding balanced predictions. Figures 6 and 7 examine their confusion matrices, revealing that both models exhibit balanced performance. Considering the performance-size trade-off, the MobileNet encoder appears to be the most promising.

7 Conclusions

This paper presents a comprehensive analysis of DELC classification, explicitly targeting the identification of Frank's Sign in ear images. Hard augmentation transformations may potentially yield lower quality results compared to soft augmentation transformations due to several factors. Firstly, hard augmentations tend to introduce more aggressive and drastic modifications to the input data, such as random flips, rotations, shifts, and changes in brightness and contrast. While these transformations increase the variability in the training data, they may also introduce noise and distortions that can hinder the model's ability to generalize effectively. The overly aggressive modifications can result in the model focusing on irrelevant or misleading features, leading to a degradation in performance.

In contrast, soft augmentation transformations adopt a more subtle and realistic approach by generating additional synthetic data samples with variations in rotation, lighting conditions, color attributes, and intensity. These variations align more closely with the natural variations in real-world scenarios, allowing the model to learn more accurate representations of the target class. The synthetic samples generated through soft augmentation techniques effectively expand the dataset and provide more diverse examples for the model to learn from, enhancing its ability to handle unseen variations and improve generalization.

Moreover, hard augmentations may introduce inconsistencies or artifacts that need to reflect the true characteristics of the diagonal earlobe crease accurately. For instance, aggressive transformations like flips or rotations may distort the shape or orientation of the earlobe crease, making it challenging for the model to learn and recognize genuine patterns. Soft augmentations, on the other hand, simulate variations that closely resemble real-world scenarios and preserve the key characteristics of the diagonal earlobe crease, aiding the model in effectively capturing and distinguishing these features.

In summary, the less aggressive and more realistic nature of soft augmentation transformations, along with their ability to generate synthetic samples that closely resemble natural variations, provide an advantage over hard augmentation transformations. Soft augmentations tend to preserve the essential features of the diagonal earlobe crease while expanding the dataset to facilitate better generalization and accuracy, leading to potentially superior results compared to hard augmentations.

Our research also focuses on the relationship between encoder size and performance, specifically for Diagonal Earlobe Crease detection. By considering encoder performance and size, we successfully address the challenges posed by complex image scenarios. Our selected encoder achieves remarkable accuracy in solving the DELC detection problem. Furthermore, lighter-weight encoders within the same backbone family (ResNet, DenseNet, VGG) generally outperform their larger counterparts. Through a performance-size trade-off analysis, we identify MobileNet as the most promising encoder, offering a slight decrease in performance but with significantly faster processing and lighter weight (3 times faster and 5.5 times lighter).

This line of research opens up numerous exciting opportunities in healthcare scenarios. Our proposal can potentially optimize diagnostic and prognostic pathways, develop personalized treatment strategies, and leverage larger datasets. Non-invasive DELC detection through earlobe image analysis emerges as a significant application. Additionally, monitoring the changes in earlobes over time can provide valuable insights into an individual's health status and enable medical professionals to identify potential risks.

References

1. Almonacid-Uribe., S., Santana., O.J., Hernández-Sosa., D., Freire-Obregón., D.: Deep learning for diagonal earlobe crease detection. In: Proceedings of the International Conference on Pattern Recognition Applications and Methods, pp. 74–81 (2023)
2. Alshazly, H., Linse, C., Barth, E., Martinetz, T.: Handcrafted versus CNN features for ear recognition. Symmetry **11**(12), 1493 (2019)
3. Ardila, D., Kiraly, A., Bharadwaj, S., Choi, B., Shetty, S.: End-to-end lung cancer screening with three-dimensional deep learning on low-dose chest computed tomography. Nat. Med. **25**, 954–961 (2019)
4. Boudoulas, K., Triposkiadis, F., Geleris, P., Boudoulas, H.: Coronary atherosclerosis: pathophysiologic basis for diagnosis and management. Prog. Cardiovasc. Dis. **58**, 676–692 (2016)
5. Buslaev, A., Iglovikov, V.I., Khvedchenya, E., Parinov, A., Druzhinin, M., Kalinin, A.A.: Albumentations: fast and flexible image augmentations. Information **11**(2), 125 (2020)
6. CDC: About Multiple Cause of Death, 1999–2020 (2022)
7. Cha, D., Pae, C., Seong, S.B., Choi, J.Y., Park, H.J.: Automated diagnosis of ear disease using ensemble deep learning with a big otoendoscopy image database. EBioMedicine **45**, 606–614 (2019)
8. Chollet, F.: Xception: Deep learning with depthwise separable convolutions. CoRR abs/1610.02357 (2016)
9. Deng, J., Dong, W., Socher, R., Li, L.J., Li, K., Fei-Fei, L.: ImageNet: a large-scale hierarchical image database. In: 2009 IEEE Conference on Computer Vision and Pattern Recognition, pp. 248–255 (2009)
10. Emeršič, Ž., Štruc, V., Peer, P.: Ear recognition: more than a survey. Neurocomputing **255**, 26–39 (2017). Bioinspired Intelligence for machine learning
11. Frank, S.: Aural sign of coronary-artery disease. IEEE Trans. Med. Imaging **289**(6), 327–328 (1973)
12. Freire-Obregón, D., Marsico, M.D., Barra, P., Lorenzo-Navarro, J., Castrillón-Santana, M.: Zero-shot ear cross-dataset transfer for person recognition on mobile devices. Pattern Recogn. Lett. **166**, 143–150 (2023)

13. Freire-Obregón, D., Rosales-Santana, K., Marín-Reyes, P.A., Penate-Sanchez, A., Lorenzo-Navarro, J., Castrillón-Santana, M.: Improving user verification in human-robot interaction from audio or image inputs through sample quality assessment. Pattern Recogn. Lett. **149**, 179–184 (2021)
14. He, K., Zhang, X., Ren, S., Sun, J.: Deep residual learning for image recognition. In: 2016 IEEE Conf. on Computer Vision and Pattern Recognition, pp. 770–778 (2016)
15. Hirano, H., Katsumata, R., Futagawa, M., Higashi, Y.: Towards view-invariant expression analysis using analytic shape manifolds. In: 2016 IEEE Engineering in Medicine and Biology Society, pp. 2374–2377 (2016)
16. Howard, A.G., et al.: MobileNets: efficient convolutional neural networks for mobile vision applications. CoRR abs/1704.04861 (2017)
17. Huang, G., Liu, Z., van der Maaten, L., Weinberger, K.Q.: Densely connected convolutional networks. CoRR abs/1608.06993 (2016)
18. Isgum, I., Prokop, M., Niemeijer, M., Viergever, M.A., van Ginneken, B.: Automatic coronary calcium scoring in low-dose chest computed tomography. IEEE Trans. Med. Imaging **31**(12), 2322–2334 (2012)
19. Kingma, D.P., Ba, J.: Adam: a method for stochastic optimization. In: 2015 International Conference on Learning Representations (2015)
20. Lee, M.C.H., Petersen, K., Pawlowski, N., Glocker, B., Schaap, M.: TeTrIS: template transformer networks for image segmentation with shape priors. IEEE Trans. Med. Imaging **38**(11), 2596–2606 (2019)
21. Mallinson, T., Brooke, D.: Limited diagnostic potential of diagonal earlobe crease. Ann. Emerg. Med. **70**(4), 602–603 (2017)
22. Pasternac, A., Sami, M.: Predictive value of the ear-crease sign in coronary artery disease. Can. Med. Assoc. J. **126**(6), 645–649 (1982)
23. Rodríguez-López, C., Garlito-Díaz, H., Madroñero-Mariscal, R., López-de Sá, E.: Earlobe crease shapes and cardiovascular events. Am. J. Cardiol. **116**(2), 286–293 (2015)
24. Sanchis-Gomar, F., Perez-Quilis, C., Leischik, R., Lucia, A.: Epidemiology of coronary heart disease and acute coronary syndrome. Ann. Transl. Medi. **4**(13), 1–12 (2016)
25. Santana, O.J., Freire-Obregón, D., Hernández-Sosa, D., Lorenzo-Navarro, J., Sánchez-Nielsen, E., Castrillón-Santana, M.: Facial expression analysis in a wild sporting environment. In: Multimedia Tools and Applications (2022)
26. Simonyan, K., Zisserman, A.: Very deep convolutional networks for large-scale image recognition. In: Bengio, Y., LeCun, Y. (eds.) 2015 International Conference on Learning Representations (2015)
27. Steinberg, A., Rosner, F.: Encyclopedia of Jewish Medical Ethics. Feldheim Publishers, Spring Valley (2003)
28. Stoyanov, G., Dzhenkov, D., Petkova, L., Sapundzhiev, N., Georgiev, S.: The histological basis of Frank's sign. Head Neck Pathol. **15**(2), 402–407 (2021)
29. Szegedy, C., Vanhoucke, V., Ioffe, S., Shlens, J., Wojna, Z.: Rethinking the inception architecture for computer vision. In: 2016 IEEE Conference on Computer Vision and Pattern Recognition, pp. 2818–2826 (2016)
30. Tsao, C., Aday, A., Almarzooq, Z., Beaton, A., Bittencourt, M., Boehme, A.: Heart Disease and Stroke Statistics-2022 Update: A Report from the American Heart Association. Circulation **145**(8), e153–e639 (2022)
31. Wang, Y., et al.: Deep learning in automated region proposal and diagnosis of chronic otitis media based on computed tomography. Ear Hear. **41**(3), 669–677 (2020)
32. Wieckowski, K.: Diagonal earlobe crease (Frank's Sign) for diagnosis of coronary artery disease: a systematic review of diagnostic test accuracy studies. J. Clin. Med. **10**(13), 2799 (2021)

33. Wolterink, J.M., Leiner, T., Takx, R.A.P., Viergever, M.A., Isgum, I.: Automatic coronary calcium scoring in non-contrast-enhanced ECG-triggered cardiac CT with ambiguity detection. IEEE Trans. Med. Imaging **34**(9), 1867–1878 (2015)
34. Yamak, D., Panse, P., Pavlicek, W., Boltz, T., Akay, M.: Non-calcified coronary atherosclerotic plaque characterization by dual energy computed tomography. IEEE J. Biomed. Health Inform. **18**(3), 939–945 (2014)
35. Zeleznik, R., et al.: Deep convolutional neural networks to predict cardiovascular risk from computed tomography. Nat. Commun. **12**(1), 715 (2021)
36. Zeng, X., et al.: Efficient and accurate identification of ear diseases using an ensemble deep learning model. Sci. Rep. **11**(1), 10839 (2021)
37. Zhao, F., Wu, B., Chen, F., He, X., Liang, J.: An automatic multi-class coronary atherosclerosis plaque detection and classification framework. Med. Biol. Eng. Comput. **57**(1), 245–257 (2019)
38. Zreik, M., van Hamersvelt, R.W., Wolterink, J.M., Leiner, T., Viergever, M.A., Isgum, I.: A recurrent CNN for automatic detection and classification of coronary artery plaque and stenosis in coronary CT angiography. IEEE Trans. Med. Imaging **38**(7), 1588–1598 (2019)

Distance Transform in Images and Connected Plane Graphs

Majid Banaeyan(✉) and Walter G. Kropatsch

Pattern Recognition and Image Processing Group, TU Wien, Vienna, Austria
{majid,krw}@prip.tuwien.ac.at

Abstract. The distance transform (DT) serves as a crucial operation in numerous image processing and pattern recognition methods, finding broad applications in areas such as skeletonization, map-matching robot self-localization, biomedical imaging, and analysis of binary images. The concept of DT computation can also be extended to non-grid structures and graphs for the calculation of the shortest paths within a graph. This paper introduces two distinct algorithms: the first calculates the DT within a connected plane graph, while the second is designed to compute the DT in a binary image. Both algorithms demonstrate parallel logarithmic complexity of $\mathcal{O}(log(n))$, with n representing the maximum diameter of the largest region in either the connected plane graph or the binary image. To attain this level of complexity, we make the assumption that a sufficient number of independent processing elements are available to facilitate massively parallel processing. Both methods utilize the hierarchical irregular pyramid structure, thereby retaining topological information across regions. These algorithms operate entirely on a local level, making them conducive to parallel implementations. The GPU implementation of these algorithms showcases high performance, with memory bandwidth posing the only significant constraint. The logarithmic complexity of the algorithms boosts execution speed, making them particularly suited to handling large images.

Keywords: Distance transform · Connected plane graph · Parallel logarithmic complexity · Irregular graph pyramids · Parallel processing

1 Introduction

The concept of distance transform (DT) [34], a cornerstone technique in pattern recognition and geometric computations, has a pivotal role in an array of methods. Its utility spans a wide spectrum of applications, including but not limited to skeletonization [31], robotic self-localization through map matching [36], image registration [19], template matching [26,33], line detection in manuscripts [21], and weather forecasting and analysis [9]. Employed primarily on binary images composed of background and foreground regions, the DT produces a new gray-scale image. In this transformed image, the intensity of each foreground pixel reflects the minimum distance to the background.

Supported by the Vienna Science and Technology Fund (WWTF), project LS19-013.

M. De Marsico et al. (Eds.): ICPRAM 2023, LNCS 14547, pp. 19–32, 2024.
https://doi.org/10.1007/978-3-031-54726-3_2

The application of distance transform extends beyond binary images and into the realm of connected plane graphs, enhancing its utility further [27]. A connected plane graph consists of vertices and edges, analogous to the pixels and adjacency relationships in an image. By applying distance transform on such a graph, each vertex obtains a value that signifies the shortest distance to a set of predefined background vertices.

This transformation on connected plane graphs finds immense value in network analysis [28] and routing applications [20]. It serves as a crucial step in path-finding algorithms, where identifying the shortest paths between nodes [18] can result in optimized routing [14, 15]. This is particularly useful in applications such as transport logistics, telecommunications routing [12], and even in social network analysis. Thus, distance transform not only offers valuable insights into image analysis but also aids in optimizing network structures and enhancing routing efficiency [11].

In the computation of the Distance Transform (DT), conventional algorithms applied to binary images [13,32,34] or connected plane graphs [18,22] typically propagate distances in a linear-time complexity, denoted as $\mathcal{O}(N)$. Here, N represents the quantity of pixels in a 2D binary image or the count of vertices in a connected plane graph.

Differently, this paper introduces an innovative approach that propagates distances exponentially, thereby computing the DT with a parallel time complexity of $\mathcal{O}(log(n))$. In this instance, n signifies the diameter of the most extensive connected component found within the binary image or the connected plane graph. In the pursuit of achieving parallel logarithmic complexity, it is imperative to note a key assumption we make. We presume that a sufficient number of independent processing units are readily accessible. This assumption is crucial to the successful execution of our proposed method, as it is heavily reliant on simultaneous processing capabilities.

Our proposed method builds upon the concepts presented in a recent publication [7], which leverages the hierarchical structure found in the irregular graph pyramid [4,8,24]. The methodology outlined in [7] focuses on computing the distance transform on a grid structure. Conversely, our paper introduces a novel approach for calculating the distance transform in a general, non-grid connected plane graph, an approach that also effectively solves the shortest path problem.

In this study, we propose two algorithms that exhibit parallel logarithmic complexity. In Sect. 3, we elaborate on the first method for computing the DT in a connected plane graph. Subsequently, Sect. 4 presents the second method for computing the DT in a binary image. Prior to that, in Sect. 2, we provide a recap of the background and definitions. Lastly, in Sect. 5, we evaluate and compare the results obtained from both methods.

2 Background and Definitions

The image pyramid is a stack of images, each presented at a progressively reduced resolution [8]. It is a methodology prevalent in various domains, including image processing [10] and pattern recognition [34]. This approach effectively encapsulates both local and global information across the different levels. The procedure starts with high-resolution data at the base level and progressively transmutes the local details into more abstract, global information as one ascends the pyramid [29].

Pyramids are essentially of two categories: regular and irregular [8]. In regular pyramids, the resolution decreases in a consistent pattern from one image to the next. However, the resolution reduction in irregular pyramids does not follow a fixed rate. A notable drawback of regular pyramids is their lack of shift invariance - even a minor deviation in the initial image can potentially induce significant alterations in the subsequent pyramid [8]. Irregular pyramids offer a solution to this issue, being data-driven hierarchical structures that inherently bypass the shift variance problem.

A digital image can be visualized as a neighborhood graph. Consider $G = (V, E)$ as the neighborhood graph representing image P. Here, V correlates to P, and E links neighboring pixels. The 4-neighborhood representation is typically favored to avoid intersection of edges between diagonal neighbors within 2×2 pixels, thus maintaining the graph's planarity, which would be compromised in an 8-connected graph.

Irregular graph pyramids [24] are a sequence of consecutively reduced graphs, each iteratively constructed from the graph below it through the selection of specific vertices and edges subsets. The pyramid's construction utilizes two fundamental operations: edge contraction and edge removal. In the case of edge contraction, an edge $e = (v, w)$ undergoes contraction, with the endpoints v and w merging and the edge itself being eliminated. Post-operation, the edges originally connected to the combined vertices become incident to the resulting singular vertex. Conversely, edge removal simply entails the removal of an edge without modifying the count of vertices or interfering with the incidence relationships of remaining edges. Throughout the pyramid, the vertices and edges that do not persist to the next level are termed *non-surviving*, while those that do make it to the subsequent level are classified as *surviving*.

A ***plane*** graph [37], is a graph embedded in a two-dimensional plane where its edges exclusively intersect at their endpoints. In the plane graph there are connected spaces between edges and vertices and every such connected area of the plane is called a *face*. A face's degree is quantified by the number of edges that enclose it. Further categorization introduces the notion of an *empty* face, specifically referring to a face that is demarcated by a cycle. For non-empty faces, traversal of the boundary necessitates multiple visits to certain vertices or edges [23]. Empty faces that encompass only a single edge are distinguished as empty *self-loops*. Considering an empty face of degree 2, it would encompass a pair of edges with identical endpoints. These parallel edges are called as *multiple* edges.

Definition 1 (Contraction Kernel (CK)). *A CK is a tree consisting of a surviving vertex as its root and some non-surviving neighbors with the constraint that every non-survivor can be part of only one CK [5].*

An edge of a CK is denoted by the directed edge and points towards the survivor.

A *Maximal Independent Vertex Set* (**MIVS**) [29] represents a collection of independent vertices within a connected plane graph. Here, independence implies that no two vertices within the set are neighbors [29]. The MIVS method employs an iterative stochastic process based on a uniformly distributed random variable [0, 1] assigned to each vertex [8]. Vertices corresponding to a local maximum of this random variable are surviving vertices, whereas their neighboring vertices are non-survivors. There may be some isolated vertices left, which will be connected to the local maximum in subsequent iterations for the construction of the independent set. The survival probability of a vertex is correlated to the size of its neighborhood [30], which influences the height of the pyramid and the number of iterations required to construct the pyramid [8]. A challenge with the MIVS approach is that the average degree of vertices tends to increase throughout the pyramid [25]. This leads to a reduction in the number of surviving vertices, subsequently decreasing the decimation ratio along the pyramid [16], which inadvertently increases the pyramid's height. To address this drawback, the *Maximum Independent Edge Set* (MIES) was introduced.

The **MIES** [16] applies the MIVS method to an *edge-graph* derived from the original graph G. This edge-graph comprises a vertex for each edge in G, with vertices in the edge graph being connected if their corresponding edges in G are incident to the same vertex. Consequently, the MIES introduces a maximal matching [17] on the initial graph vertices. A matching on a graph refers to a subset of its edges wherein no two edges share a common vertex. Such a matching is deemed *maximal* if it is impossible to add any additional edge without breaching the matching condition [16]. *Maximum* matching, on the other hand, represents the matching scenario that encompasses the largest possible number of edges for a given graph [16].

3 DT in a Connected Plane Graph

In a planar graph $G = (V, E)$, where V and E represent the vertices and edges respectively, distances are determined by the shortest path length [35]. Each vertex $v \in V$ has a set of neighbors denoted by $\mathcal{N}(v) = \{v\} \cup \{w \in V | e = (v, w) \in E\}$ linked via edges. Partition the vertices $V = B \cup F$ with $B \cap F = \emptyset$ into background and foreground vertices in the graph.

In the process of computing the distance transform (DT) on the graph, vertices in the set $b \in B$ act as *seed* vertices, with their respective distances initialized to zero, denoted as $DT(b) = 0$. The foreground $f \in F$ have their distances initialized to infinity, or $DT(f) = \infty$. The calculation of the distance transform utilizes an irregular graph pyramid structure [3,7]. Within this hierarchical structure, information is propagated in two primary directions: (1) **bottom-up** and (2) **top-down**. The bottom-up construction involves computing the DT for a subset of vertices, referred to as surviving vertices, progressively up the pyramid until its apex is reached. Conversely, the top-down process computes the DT for the remaining vertices, proceeding downwards until the DT for all vertices has been calculated at the base level.

3.1 Bottom-Up Construction in the Irregular Pyramid

The irregular pyramid is constructed from the original graph, G, which forms the base of the pyramid, through a bottom-up procedure. This process comprises four iterative steps at each level of the pyramid: (1) propagation of distances, (2) selection of Contraction Kernels (CKs), (3) contraction of the selected CKs, and, (4) simplification.

Distance Propagation. At each level of the pyramid, vertices with previously computed Distance Transform (DT) propagate their distances to their adjacent vertices. This propagation is described by the following relation:

$$D(v) = \min\{D(v), D(v_j) + 1 + d_c |\ v_j \in \mathcal{N}(v)\} \quad v \in F \tag{1}$$

Here, d_c is the shortest length of the cumulative lengths of the subpaths with the same end vertices resulting from several steps of contractions.

Selection of Contraction Kernels and Contraction Process. The selection of Contraction Kernels (CKs) is a crucial step in the construction of the pyramid. To ensure a fully parallel construction of the pyramid with logarithmic height, it is necessary for the selected CKs to be independent of each other [6]. To achieve this, we employ the Maximum Independent Edge Set (MIES) method [25], where the chosen CKs share no common vertices. Once the CKs are selected, they are contracted, resulting in a smaller, reduced graph at the subsequent higher level of the pyramid. The *reduction* function [8] computes the lengths of the newly contracted edges and the shortest distance of the surviving vertices.

Simplification. The contraction of selected independent Contraction Kernels (CKs) yields a smaller graph, which might contain parallel edges or empty self-loops. Having no topological information [5] these superfluous edges are referred to as *redundant* edges [2,5]. To simplify the resulting graph, these redundant edges are removed. Only the shortest of a set of multiple edges is kept.

The pyramid's construction concludes when all surviving vertices have received their corresponding DT. This graph represents the apex of the pyramid. Figure 1 provides an example of computing the DT via the bottom-up procedure. The pyramid's base level (Fig. 1a) is the original connected plane graph, which comprises a single filled vertex as the background and the remaining unfilled vertices as the foreground. During the initialization step, the background vertex and its adjacent vertices receive their DT. The edges incident to the background are colored in red, while the remaining foreground edges are shown in dark blue. The selected CKs are indicated by black arrows at various pyramid levels (Fig. 1b, d, f). Post-contraction, the resulting graph (Fig. 1c, e) contains redundant edges. For instance, in Fig. 1c, an empty self-loop is represented by "s". The edges labeled b, c, e, g, l, n, and o are redundant parallel edges. The cumulative sum of previous contractions, d_c, as denoted in Eq. (1), is illustrated by $\boxed{\text{i}}$, where $i = 1, 2$ beside vertices and their corresponding incident edges.

3.2 Top-Down Propagation

The computation of the Distance Transform (DT) for the remaining vertices with unknown distances involves two main steps, executed level by level from the apex to the base of the pyramid: (1) distance propagation, and (2) correction.

Distance Propagation. In the top-down procedure, the hierarchical structure of the pyramid is reconstructed from the apex down to the base level. This reconstruction involves the re-insertion of edges to re-establish connectivity and expand the graph. To compute the DT, each vertex with a previously computed DT propagates its distance to

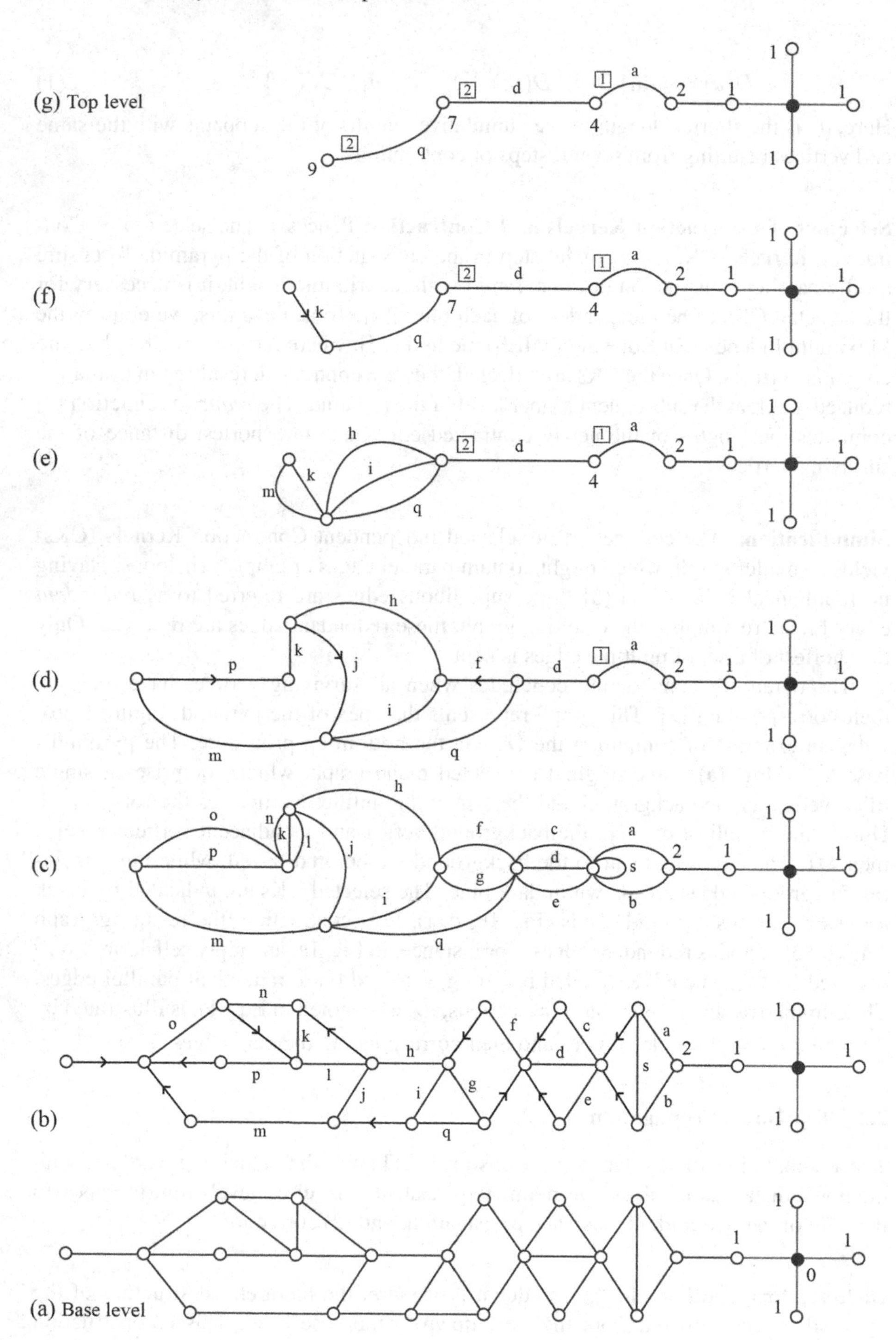

Fig. 1. Bottom-up construction in the pyramid.

its corresponding vertex at the lower level. Following this, one step of DT propagation (as described in Eq. (1)) is conducted, leading to the propagation of distances to the newly inserted vertices. This process continues until the pyramid reaches the base level, at which point all vertices possess their respective DTs.

Correction. According to Eq. (1), the difference between the DT values of any two adjacent vertices at a given pyramid level should be at most $1 + d_c$. However, during the top-down reconstruction of the pyramid, the insertion of a new edge might connect two vertices whose distance apart is greater than $1 + d_c$. In such situations, another application of Eq. (1) is needed to update the new DT and ensure accuracy of the final result.

Figure 2 illustrates the top-down reconstruction of the pyramid, with the correction of certain distances represented by bevelled red lines. Algorithm 1 demonstrates the computation of the Distance Transform (DT) in a connected plane graph.

Algorithm 1. Computing the Distance Transform (DT) in a Connected Plane Graph.

1: **Input:** Connected Plane Graph: $G = (V, E) = (B \cup F, E)$ L=pyramid's level
2: **Initialization:** Set $DT(b) = 0; \forall b \in B$, and $DT(f) = \infty$, $\forall f \in F$
3: **While** there are edges to be contracted, perform the following steps for **bottom-up** construction of the pyramid:
4: Propagate the distances using Equ. (1)
5: Select the Contraction Kernels (CKs)
6: Contract the CKs
7: $L \rightarrow L + 1$
8: Record the number of contractions
9: Simplify the resulting reduced graph
10: **End While** (Top of the pyramid is reached)
11: **While** there are vertices with unknown DT in the level below, perform the following steps for **top-down** propagation of distances:
12: $L \rightarrow L - 1$
13: Inherit the computed DT from higher levels
14: Propagate the distances using Equ. (1)
15: Implement corrections if necessary
16: **End While**

4 Distance Transform (DT) in a Binary Image

A binary image can be conceptualized as a two-dimensional matrix, with each element assuming a value of either zero or one. The neighborhood graph corresponding to this binary image, denoted by $G = (V, E)$, comprises vertices V representing pixels ($p \in P$) of the image, and edges E signifying the adjacency relationships between these pixels. For the purpose of creating a plane neighborhood graph, we assume 4-nearest neighbor relations between the pixels. The reason is that the 8-connectivity would not be a plane graph [23].

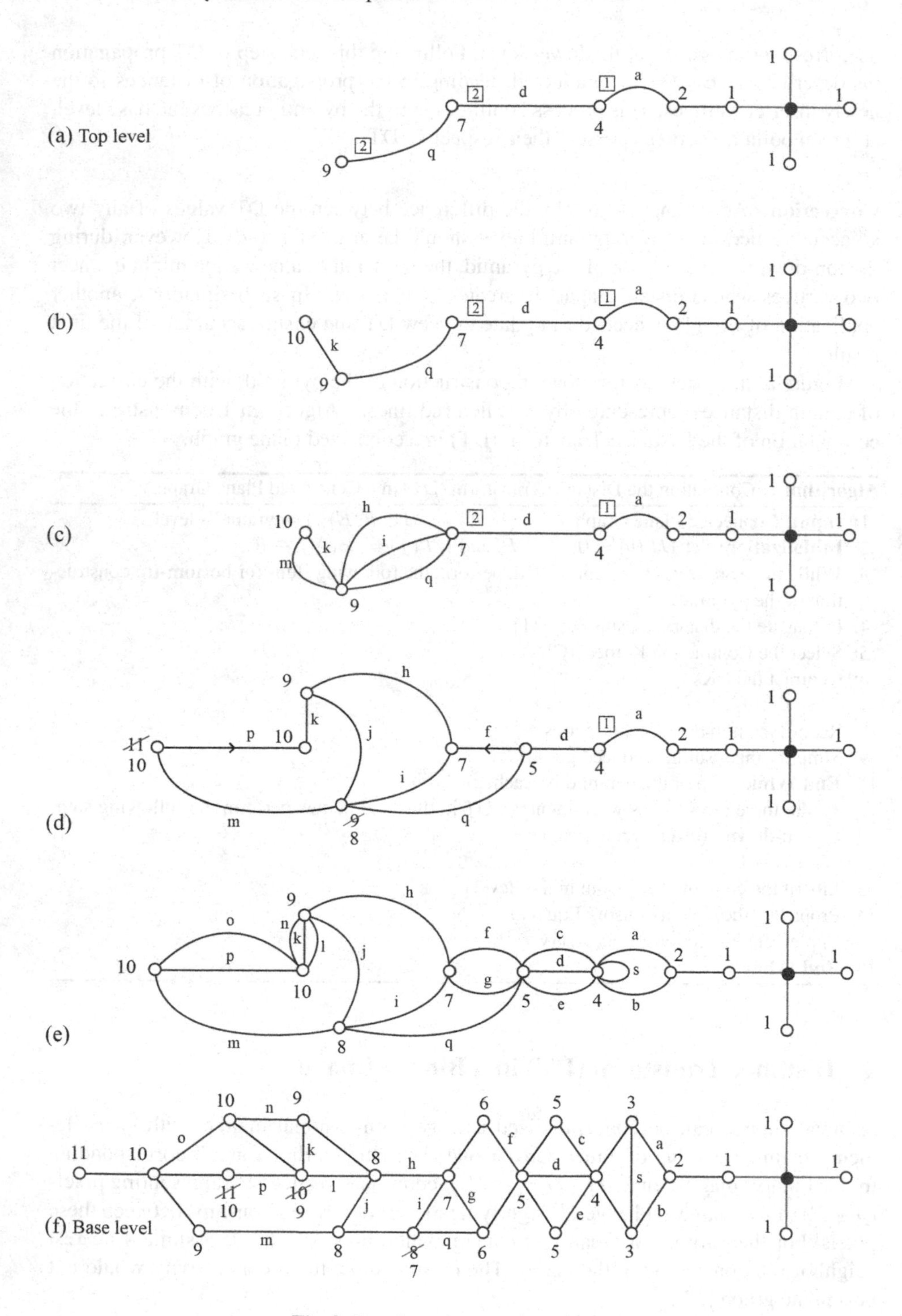

Fig. 2. Top-down propagation of DT.

The computation of the DT in the binary image employs the use of an irregular pyramid as proposed by Banaeyan (2023) [7]. In this hierarchical structure, an efficient selection of Contraction Kernels (CKs) is enabled by defining a *total order* through the indices of the vertices, as suggested in [3].

Suppose the binary image comprises M rows and N columns, with pixel $(1,1)$ situated at the upper-left corner and pixel (M,N) at the lower-right corner. In such a setup, the vertices of the corresponding graph G are assigned a unique index as follows, based on the approach proposed in [5]:

$$Idx : [1, M] \times [1, N] \rightarrow [1, M \cdot N] \subset \mathbb{N}, \quad Idx(r, c) = (c - 1) \cdot M + r \tag{2}$$

where r and c correspond to the row and column of the pixel, respectively.

The Distance Transform (DT) in the binary image can be calculated using a method similar to Algorithm 1. However, two modifications are incorporated. Firstly, in contrast to computing the DT in the connected plane graph, the simplification step is executed before the contraction by Contraction Kernels (CKs), which expedites the pyramid construction. Secondly, by knowing the coordinates of the vertices, the propagation formula can be directly computed based on the indices of the coordinates, as proposed in [5]:

$$D(v_i) = \min\{D(v_i), D(v_j) + \begin{cases} 1 & if\, T = 1 \\ \frac{T}{M} & if\, T \neq 1 \end{cases} \} \tag{3}$$

where

$$v_j \in \mathcal{N}(v_i), \quad T = |Idx(v_i) - Idx(v_j)| \tag{4}$$

Algorithm 2 outlines the specifics of the proposed method.

Algorithm 2. Computing the Distance Transform (DT) in a binary image.

Input: Neighborhood graph of a binary image: $G = (V, E)$, L=pyramid's level
2: **Initialization:** Set $DT(b) = 0; \forall b \in B$, and $DT(f) = \infty, ; \forall f \in F$
While there are edges to be contracted, perform the following steps for **bottom-up** construction of the pyramid:
4: Propagate the distances using Equ. (1)
Select the Contraction Kernels (CKs)
6: Identify the Redundant Edges
Remove the Redundant Edges
8: Contract the CKs
$L \rightarrow L + 1$
10: **End While** (Top of the pyramid is reached)
While there are vertices with unknown DT in the level below, perform the following steps for **top-down** propagation of distances:
12: $L \rightarrow L - 1$
Inherit the computed DT from higher levels
14: Propagate the distances using Equ. (1)
Implement corrections if necessary
16: **End While**

It is noteworthy to mention that in calculating the DT of the binary image, the selection of Contraction Kernels (CKs) is executed differently. While in the plane graph the

Maximum Independent Edge Set (MIES) method selects independent CKs at each pyramid level, the binary image employs a different approach. Here, equivalent contraction kernels (ECKs) [5] are selected initially, followed by a logarithmic encoding process which divides the ECKs into a set of independent edges.

5 Evaluation and Results

To underscore the benefits of the introduced logarithmic algorithm, we conducted performance analyses comparing its execution times to those of two other GPU-based methods, MeijsterGPU and FastGPU, as presented in [1]. The experimental environment consisted of MATLAB software running on an AMD Ryzen 7 2700X, 3.7GHz CPU, and an NVIDIA GeForce GTX 2080 TI GPU. The experiments comprised three distinct types of images: Random (Ran.), Mitochondria (Mit.), and MRI. Table 1 elucidates the results obtained from these experimental scenarios. Table 1 [7] presents the image size in its first column, followed by execution times (in ms) for our proposed Logarithmic DT (Log DT) algorithm across three distinct image types, and concludes with the execution times of two other methods. Figure 3 provides a visual comparison of the logarithmic algorithm's performance across the image types. As the Random images have smaller foreground elements compared to the other types, they are processed more rapidly. In Fig. 4, the logarithmic DT's performance is contrasted with that of the MeijsterGPU and FastGPU methods. Our proposed algorithm not only outpaces the others but also displays superior handling of larger images.

Table 1. Execution (ms) of proposed Logarithmic DT, MeijsterGPU and FastGPU [7].

Image-size	Mit. Log DT	MRI Log DT	Ran. Log DT	Ran. MeijsterGPU	Ran. FastGPU
256×256	0.0953	0.1209	0.0645	3.8524	1.7844
512×512	0.410	0.7310	0.3152	14.2005	4.2309
1024×1024	2.6308	5.1501	0.9364	25.8475	12.4668
2048×2048	4.1088	8.9506	1.8577	110.7817	44.9560

It is worth highlighting that all operations within the proposed algorithms are independent and localized, allowing each GPU thread within the shared memory to be allocated to a distinct local process. This distribution creates a performance bottleneck, determined by the capacity of the shared memory. Nevertheless, with an ample number of independent processing elements, the algorithms can operate in full parallelism and maintain logarithmic complexity.

Currently, our team is engaged in the "Water's Gateway to Heaven" project[1], which focuses on high-resolution X-ray micro-tomography (μCT) and fluorescence microscopy. The 3D images involved in this project have dimensions exceeding 2000 in each direction, and the DT plays a crucial role in the challenging task of cell separation. We anticipate that our proposed method will significantly enhance the computational efficiency required for processing such large-scale 3D images. As part of our future work, we plan to evaluate the effectiveness of our method on diverse graph datasets.

[1] https://waters-gateway.boku.ac.at/.

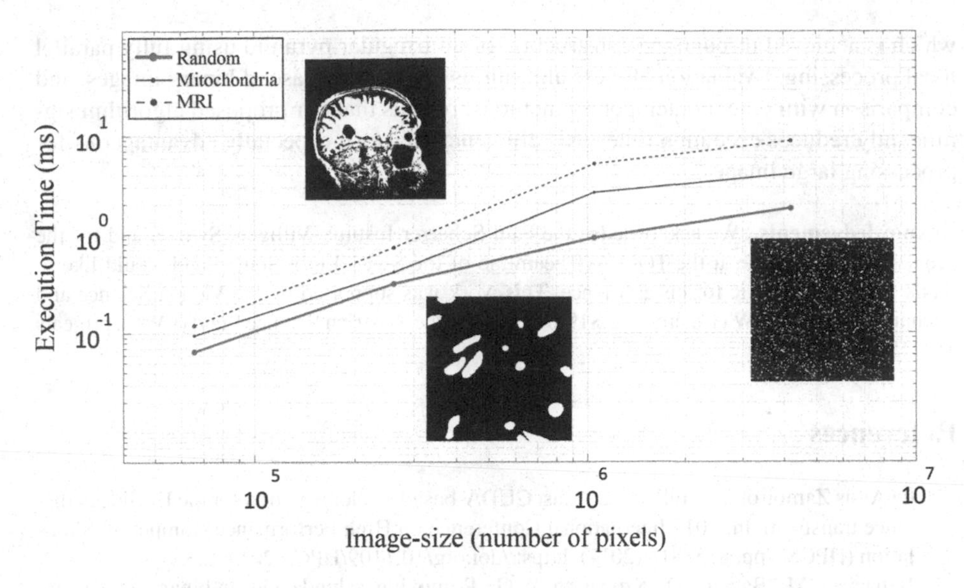

Fig. 3. The proposed logarithmic DT over different images [7].

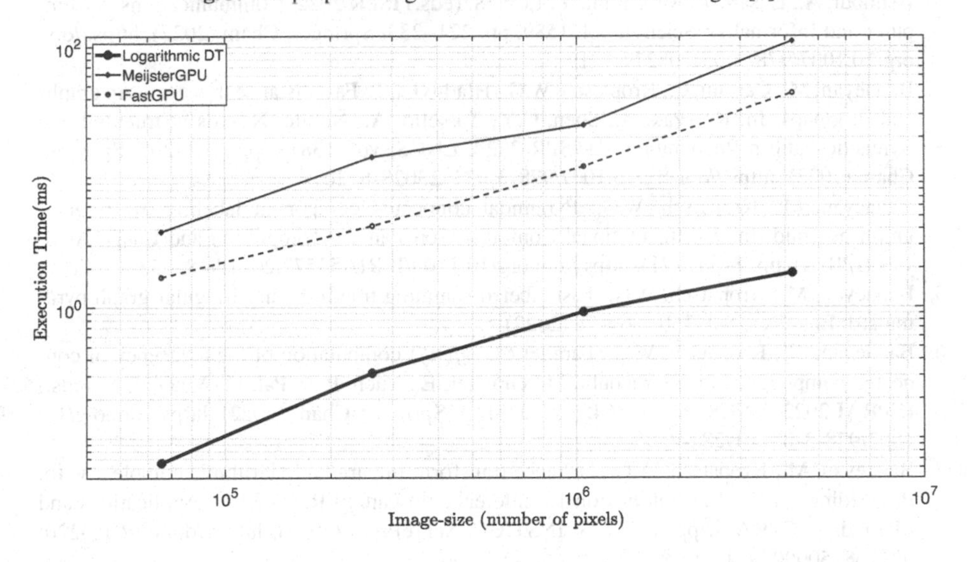

Fig. 4. Comparison of the proposed algorithm with MeijsterGPU and FastGPU [1,7].

6 Conclusion

The distance transform (DT) calculates the shortest distance to reach a point from the set of initial points. In this study, we expanded the concept of DT computation from images (grid structures) to connected plane graphs (non-grid structures). This expansion can enhance the efficiency of solving the shortest path problem in graph theory. The primary advantage of the proposed algorithms lies in their parallel logarithmic complexity, which is achieved through the construction of an irregular pyramid using fully parallel local processing. Evaluation of these algorithms using a database of binary images, and comparison with other contemporary methods, reveals that our proposed algorithm significantly reduces execution time. This efficiency makes it especially advantageous for processing large images.

Acknowledgements. We acknowledge the Paul Scherrer Institut, Villigen, Switzerland for the provision of beamtime at the TOMCAT beamline of the Swiss Light Source and would like to thank Dr. Goran Lovric for his assistance. This work was supported by the Vienna Science and Technology Fund (WWTF), project LS19-013, and by the Austrian Science Fund (FWF), projects M2245 and P30275.

References

1. de Assis Zampirolli, F., Filipe, L.: A fast CUDA-based implementation for the Euclidean distance transform. In: 2017 International Conference on High Performance Computing Simulation (HPCS), pp. 815–818 (2017). https://doi.org/10.1109/HPCS.2017.123
2. Banaeyan, M., Batavia, D., Kropatsch, W.G.: Removing redundancies in binary images. In: Bennour, A., Ensari, T., Kessentini, Y., Eom, S. (eds.) ISPR 2022. Communications in Computer and Information Science, vol. 1589, pp. 221–233. Springer, Cham (2022). https://doi.org/10.1007/978-3-031-08277-1_19
3. Banaeyan, M., Carratù, C., Kropatsch, W.G., Hladůvka, J.: Fast distance transforms in graphs and in gmaps. In: Krzyzak, A., Suen, C.Y., Torsello, A., Nobile, N. (eds.) Structural and Syntactic Pattern Recognition. S+SSPR 2022. LNCS, vol. 13813. pp. 193–202. Springer, Cham (2022). https://doi.org/10.1007/978-3-031-23028-8_20
4. Banaeyan, M., Kropatsch, W.G.: Pyramidal connected component labeling by irregular graph pyramid. In: 5th International Conference on Pattern Recognition and Image Analysis (IPRIA), pp. 1–5 (2021). https://doi.org/10.1109/IPRIA53572.2021.9483533
5. Banaeyan, M., Kropatsch, W.G.: Fast labeled spanning tree in binary irregular graph pyramids. J. Eng. Res. Sci. **1**(10), 69–78 (2022)
6. Banaeyan, M., Kropatsch, W.G.: Parallel $\mathcal{O}(log(n))$ computation of the adjacency of connected components. In: El Yacoubi, M., Granger, E., Yuen, P.C., Pal, U., Vincent, N. (eds.) ICPRAI 2022. LNCS, vol. 13364, pp. 102–113. Springer, Cham (2022). https://doi.org/10.1007/978-3-031-09282-4_9
7. Banaeyan, M., Kropatsch, W.G.: Distance transform in parallel logarithmic complexity. In: Proceedings of the 12th International Conference on Pattern Recognition Applications and Methods - ICPRAM, pp. 115–123. INSTICC, SciTePress (2023). https://doi.org/10.5220/0011681500003411
8. Brun, L., Kropatsch, W.G.: Hierarchical graph encodings. In: Lézoray, O., Grady, L. (eds.) Image Processing and Analysis with Graphs: Theory and Practice, pp. 305–349. CRC Press (2012)
9. Brunet, D., Sills, D.: A generalized distance transform: theory and applications to weather analysis and forecasting. IEEE Trans. Geosci. Remote Sens. **55**(3), 1752–1764 (2017). https://doi.org/10.1109/TGRS.2016.2632042
10. Burt, P.J., Hong, T.H., Rosenfeld, A.: Segmentation and estimation of image region properties through cooperative hierarchial computation. IEEE Trans. Syst. Man Cybern. **11**(12), 802–809 (1981)

11. Demaine, E.D., Hajiaghayi, M., Klein, P.N.: Node-weighted Steiner tree and group Steiner tree in planar graphs. In: Albers, S., Marchetti-Spaccamela, A., Matias, Y., Nikoletseas, S., Thomas, W. (eds.) ICALP 2009. LNCS, vol. 5555, pp. 328–340. Springer, Heidelberg (2009). https://doi.org/10.1007/978-3-642-02927-1_28
12. Ellinas, G., Stern, T.E.: Automatic protection switching for link failures in optical networks with bi-directional links. In: Proceedings of GLOBECOM'96. 1996 IEEE Global Telecommunications Conference, vol. 1, pp. 152–156. IEEE (1996)
13. Fabbri, R., Costa, L.D.F., Torelli, J.C., Bruno, O.M.: 2D Euclidean distance transform algorithms: a comparative survey. ACM Comput. Surv. (CSUR) **40**(1), 1–44 (2008)
14. Frederickson, G.N.: Fast algorithms for shortest paths in planar graphs, with applications. SIAM J. Comput. **16**, 1004–1022 (1987)
15. Frey, H.: Scalable geographic routing algorithms for wireless ad hoc networks. IEEE Netw. **18**(4), 18–22 (2004)
16. Haxhimusa, Y.: The Structurally Optimal Dual Graph Pyramid and Its Application in Image Partitioning, vol. 308. IOS Press, Amsterdam (2007)
17. Haxhimusa, Y., Glantz, R., Kropatsch, W.G.: Constructing stochastic pyramids by MIDES - maximal independent directed edge set. In: Hancock, E., Vento, M. (eds.) 4th IAPR-TC15 Workshop on Graph-based Representation in Pattern Recognition, vol. 2726, pp. 24–34. Springer, Heidelberg (2003). https://doi.org/10.1007/3-540-45028-9_3. http://www.prip.tuwien.ac.at/people/krw/more/papers/2003/GbR/haxhimusa.pdf
18. Henzinger, M.R., Klein, P., Rao, S., Subramanian, S.: Faster shortest-path algorithms for planar graphs. J. Comput. Syst. Sci. **55**(1), 3–23 (1997). https://doi.org/10.1006/jcss.1997.1493. https://www.sciencedirect.com/science/article/pii/S0022000097914938
19. Hill, B., Baldock, R.A.: Constrained distance transforms for spatial atlas registration. BMC Bioinform. **16**(1), 1–10 (2015)
20. Hong, S.H., Tokuyama, T.: Beyond Planar Graphs. In: Communications of NII Shonan Meetings, vol. 1, pp. 11–29. Springer, Singapore (2020). https://doi.org/10.1007/978-981-15-6533-5
21. Kassis, M., El-Sana, J.: Learning free line detection in manuscripts using distance transform graph. In: 2019 International Conference on Document Analysis and Recognition (ICDAR), pp. 222–227 (2019)
22. Klein, P.N., Mozes, S., Sommer, C.: Structured recursive separator decompositions for planar graphs in linear time. In: Symposium on the Theory of Computing (2012)
23. Klette, R.: Concise Computer Vision, vol. 233. Springer, London (2014). https://doi.org/10.1007/978-1-4471-6320-6
24. Kropatsch, W.G.: Building irregular pyramids by dual graph contraction. IEE-Proc. Vis. Image Sig. Process. **142**(6), 366–374 (1995)
25. Kropatsch, W.G., Haxhimusa, Y., Pizlo, Z., Langs, G.: Vision pyramids that do not grow too high. Pattern Recogn. Lett. **26**(3), 319–337 (2005)
26. Lindblad, J., Sladoje, N.: Linear time distances between fuzzy sets with applications to pattern matching and classification. IEEE Trans. Image Process. **23**(1), 126–136 (2014). https://doi.org/10.1109/TIP.2013.2286904
27. Lotufo, R., Falcao, A., Zampirolli, F.: Fast euclidean distance transform using a graph-search algorithm. In: Proceedings 13th Brazilian Symposium on Computer Graphics and Image Processing (Cat. No.PR00878), pp. 269–275 (2000). https://doi.org/10.1109/SIBGRA.2000.883922
28. Masucci, A.P., Smith, D., Crooks, A., Batty, M.: Random planar graphs and the London street network. Eur. Phys. J. B **71**, 259–271 (2009)
29. Meer, P.: Stochastic image pyramids. Comput. Vis. Graph. Image Process. **45**(3), 269–294 (1989)

30. Montanvert, A., Meer, P., Rosenfeld, A.: Hierarchical image analysis using irregular tessellations. In: Faugeras, O. (ed.) ECCV 1990. LNCS, vol. 427, pp. 28–32. Springer, Heidelberg (1990). https://doi.org/10.1007/BFb0014847
31. Niblack, C., Gibbons, P.B., Capson, D.W.: Generating skeletons and centerlines from the distance transform. CVGIP: Graph. Models Image Process. **54**(5), 420–437 (1992)
32. Nilsson, O., Söderström, A.: Euclidian distance transform algorithms: a comparative study (2007)
33. Prakash, S., Jayaraman, U., Gupta, P.: Ear localization from side face images using distance transform and template matching. In: 2008 First Workshops on Image Processing Theory, Tools and Applications, pp. 1–8. IEEE (2008)
34. Rosenfeld, A., Pfaltz, J.L.: Sequential operations in digital picture processing. Assoc. Comput. Mach. **13**(4), 471–494 (1966)
35. Sharaiha, Y.M., Christofides, N.: A graph-theoretic approach to distance transformations. Pattern Recogn. Lett. **15**(10), 1035–1041 (1994). https://doi.org/10.1016/0167-8655(94)90036-1. https://www.sciencedirect.com/science/article/pii/0167865594900361
36. Sobreira, H., et al.: Map-matching algorithms for robot self-localization: a comparison between perfect match, iterative closest point and normal distributions transform. J. Intell. Robot. Syst. **93**(3), 533–546 (2019)
37. Trudeau, R.: Introduction to Graph Theory. Dover Books on Mathematics, Dover Pub (1993)

Real-World Indoor Location Assessment with Unmodified RFID Antennas

Pedro Sobral[1,2], Rui Santos[1], Ricardo Alexandre[2], Pedro Marques[1], Mário Antunes[1,2(✉)], João Paulo Barraca[1,2], João Silva[3], and Nuno Ferreira[3]

[1] Departamento de Eletrónica, Telecomunicações e Informática, Universidade de Aveiro, Aveiro, Portugal
{sobral,ruipedro99,pedromm,mario.antunes,jpbarraca}@ua.pt

[2] Instituto de Telecomunicações, Universidade de Aveiro, Aveiro, Portugal
{sobral,rjfae,mario.antunes,jpbarraca}@av.it.pt

[3] Think Digital, Aveiro, Portugal
{jsilva,nferreira}@thinkdigital.pt

Abstract. The management of health systems has been one of the main challenges in several countries, especially where the aging population is increasing. This led to the adoption of smarter technologies as means to automate, and optimize processes within hospitals. One of the technologies adopted is active location tracking, which allows the staff within the hospital to quickly locate any sort of entity, from key persons to patients or equipment. In this work, we focus on exploring ML models to develop a reliable method for active indoor location tracking based on off the shelf RFID antennas with UHF passive tags. The presented work describes the full development of the solution, from the initial development made within a controlled environment, to the final evaluation made on a real health clinic. The proposed solution was able to achieved 0.47 m on average on a complex medical environment, with unmodified hardware.

Keywords: Indoor location · Machine learning · Passive RFID tag · Regression models

1 Introduction

A fundamental governance concern is the administration of healthcare systems and their economic viability, particularly in nations dealing with an increasingly sizable ageing population. Hospitals and clinical centres are actively looking for novel approaches to improve service efficiency, cut costs, and improve patient happiness due to the soaring patient numbers and the difficulties in matching annual budgets with this expansion. Some hospitals are starting to use active localization technology like Radio Frequency Identification (RFID) Radio Frequency Identification (RFID) [6,11] to expedite these procedures. Utilising Radio Frequency (RF) technologies makes it possible to track the location and use of medical equipment, keep tabs on the inventory and distribution of patient prescriptions, and even keep track of patient mobility across hospital departments.

M. De Marsico et al. (Eds.): ICPRAM 2023, LNCS 14547, pp. 33–45, 2024.
https://doi.org/10.1007/978-3-031-54726-3_3

The main goal of this study was to look into various approaches for indoor resource localization using Ultra High Frequency (UHF) passive tags while sticking to certain restrictions. Rapid response times, pinpoint location accuracy, minimal data requirements for each prediction (to reduce power consumption by components of the RFID radios), and, most importantly, compatibility with readily available, unaltered off-the-shelf hardware were all requirements that the models developed in this research had to meet. This study is a component of a larger initiative to create an inexpensive indoor positioning system using widely available RFID antennas and tags.

In this work, we studied how Machine Learning (ML) models can be used to predict asset locations by utilising passive tags. Our study expands on earlier study that examined related hardware in a controlled environment and produced encouraging results [8]. The main innovation in this study focuses on the evaluation of the previous approach using actual data acquired from a medical clinic. In addition, we have added improvements to the system's functionality that have been verified through practical testing in a clinical setting.

The remaining document is organized as follows. Section 2 described the current state of the art for active indoor location. The following section (Sect. 3) describes the hardware used in the execution of this study. Section 4 presents the previous proposed solution, that serves as the basis for this work. In Sect. 5 we describe the real-word scenario where the antennas where deployed. The new models (and the main contribution of this work) are presented in Sect. 6. Finally, the conclusion can be found on Sect. 7.

2 State of the Art

Due to their wide range of technological capabilities and the significant value they provide to numerous business sectors, Indoor Positioning System (IPS) have seen a rise in popularity over the previous two decades.

There are multiple techniques used in location systems, such as multilateration [1], angulation [7], fingerprinting [10] and others. These techniques require some information provided by the antennas and tags used, like the Time of Arrival (ToA) [9], Angle of Arrival (AoA) [15] and Received Signal Strength Indication (RSSI) [4]. The tags used in these systems can be active or passive, but they often require unique design and are not widely available in products from the main stream market. They are mostly present in specialised products and research projects. We will go into more detail about the most pertinent works that offer remedies for situations that are similar in the parts that follow.

SpotON [4] is a location system that uses RSSI to locate active RFID tags in a three-dimensional space. LANDMARC [5] is a system that also reflects the relationship between RSSI and power levels, and makes use of reference tags and the K-NN algorithm to estimate positions. Results show an accuracy of 2 m and a location delay of 7.5 s. In [10] Dwi *et al.* propose a fingerprinting based positioning system using a Random Forest (RF) algorithm and RSSI data, which achieved an error of 0.5 m, which is 18% lower than the compared Euclidean distance method. In [2], Lummanee et al. compare the performance of a Gradient Boosting algorithm to a typical Decision Tree (DT) applied in a positioning system. The experiment was based on a 324 m^2 area divided in 9 zones. The DT based Gradient Boosting algorithm achieved an estimation error of 0.754 m for 19 reference radio signals at 50 samples per zone, 17.8% more accurate than the typical DT. In [3] Jae *et al.* developed a passive RFID based localization system which uses RSSI information and reference tags to predict one-dimensional position of the asset. It achieves an error of 0.2089 m using the K-NN technique in a 3 m space.

These methods are mainly based on RSSI, which has the disadvantage of suffering greatly from attenuation due to internal obstacles and dynamic environments. Unlike SpotON and LANDMARC, the approach in [14] by Wilson *et al.* does not depend on RSSI, however, is based on the same RSSI principles. This research work is based on passive RSSI technology. Two scenarios of stationary and mobile RSSI tags are considered. The method gives tag count percentages for various signal attenuation conditions. The tags are located by recording characteristic curves of readings under different attenuation values at multiple locations in an environment. Similarly, Vorst et al. [12] use passive RFID tags and an onboard reader to locate mobile objects. Particle Filter (PF) technique is exploited to estimate the location from a prior learned probabilistic model, achieving a precision of 0.20–0.26 m.

By carefully designing and validating ML models, we in our work take into consideration these earlier research initiatives and increase the precision of our earlier work.

3 RFID Antenna Description

Given the environment where this work has developed, the hardware was pre-selected (as seen in Fig. 1). The antenna part is composed of two units (processing + radio), that communicate with each other through a physical bus (RS232, RS485 or Ethernet). The local processing unit was designed to communicate through Long Term Evolution (LTE) and Ethernet, exposing the radio frontend to the backend systems. It is powered through a 230 V Alternating Current (AC) power supply and uses an Advanced RISC Machine (ARM) processor running a GNU/Linux operating system (for low power consumption).

The antenna model used in our system is pretty typical and is used exactly as the manufacturer intended it to be, without any special firmware alterations. This decision lowers the cost of the system while also improving usability and accessibility. But there is a catch: the antenna's processing power or the data it transmits might not be at their best. The vendor claims that the firmware level automatically compensates for gain in order to aid in the detection of passive tags. However, this compensation has a substantial impact on how well it fits our planned scenario. Our work's main goal is to offer value by providing an effective indoor positioning solution, even when using unmodified hardware. It is important to keep in mind that, in the hypothetical situation, more reliable alternatives utilising active tags quickly exceed the average cost per patient, making them unworkable.

The communication between the antenna and the tags is made through a carrier wave in the 865–868 MHz (UHF) frequency range as defined in the EN 302 208 v3.2.0[1] directive for the European region, and cannot exceed 2 W emission power. In this way,

[1] https://www.etsi.org/deliver/etsi_en/302200_302299/302208/03.02.00_20/en_302208v030200a.pdf.

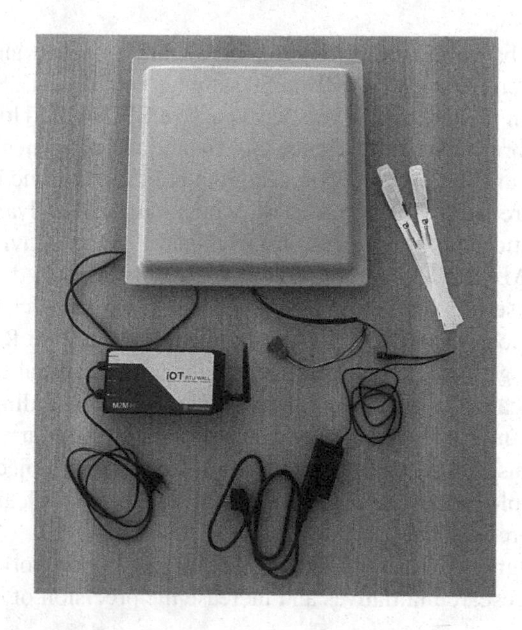

Fig. 1. Smart Antenna and passive RFID tags used for data gathering [8].

the antenna controller allows the RF emission power adjustment 0–300 mW, allowing readings up to 25 m and writings up to 6 m according to the manufacturer. The antenna polarization is circular with a gain of 12 dBi. The controller uses the Impinj R2000 chipset supporting the EPC C1 GEN2 protocol[2], ISO18000-6C[3] (see Table 1). This setup should be one of the most commonly used, as the hardware and chipset are commonly used for similar tasks. We see this as a major contribution from our work, as the output can be applied to a wide set of existing or future, deployments.

The real-time communication with the smart antenna is achieved using an MQTT broker, over which we implemented a key set of control functions: a) Definition of emission power of the antenna; b) Tag reading request over a time window (burst); c) Return data obtained at the end of the reading; d) Direct interaction with an antenna to manage it.

4 Previous Solution

This works expands on a previous one [8], which explored the same hardware but on a controlled environment. In this section we present a brief summary of the results achieved from our previous solution, which we expanded in this new work. Both works share the same environment, requirements, and the approach of using ML to provide usefulness to off the shelf hardware.

To achieve the previous solution we run three experiments to determine the best input features and ML model for location prediction.

[2] https://www.gs1.org/standards/rfid/uhf-air-interface-protocol.

[3] https://www.iso.org/standard/59644.html.

Table 1. Specification of the RFID UHF reader and writer [8].

Product Parameter	Parameter Description
Model	ACM818A UHF (20M)
Tag Protocol	EPC C1 GEN2 \ ISO 18000-6C
Output Power	Step interval 1.0 dB, maximum + 30 dBm
RF Power Output	0.1W–1W
Built-in Antenna	12dbi linear polarization
Type	antenna
Communication Ports	1) RS-232 2) RS-485 3) Wiegand 26 \ 32 bits
Communication Rates	115200bps
Reading/Writing	20 m
Multi-tags Reading	200tags/s
Working Voltage	DC +12 V

Regarding the input features, we considered two different ones, apart from the typical RSSI, the number of tag activation's and the average time between activation's were also considered. Our results showed that these input features were robust and improved greatly the precision of the location solution.

Furthermore, we explored the usage of anchor points (static tags with well known location) for model evaluation and continuously training in a dynamic environment.

Finally, we were able to establish the baseline values to the models precision on a controlled environment. Our results were aligned with the state of the art, the solution obtained an error of 0.00 m within a range of 5 m and an error of 0.55 m within a range of 10 m, resulting in an average error of 0.275 m.

5 Scenario

As previously mentioned, the previous solution was developed based on three experiments conducted in a controlled environment. This works is aligned with a National Project, as such we had access to a real health center were five antennas, and eighteen anchor were deployed. Figure 2) depicts the location of the antennas and anchors in our real-world scenario.

With this placement, the next step was to choose the best powers to generate our dataset. We want to maximize the model precision, but also wanted to increase the maximum number of tags detected. Due to latency and processing constrains we are limited to query with two power levels. After emitting a signal at the specified power level, we wait three seconds for tags responses. Querying more than two power levels would increase the latency of the system, which would impact the remaining components. Moreover, the connection system will need to store all tag activations detected during the scanning period, which presents a processing, memory and communication burden.

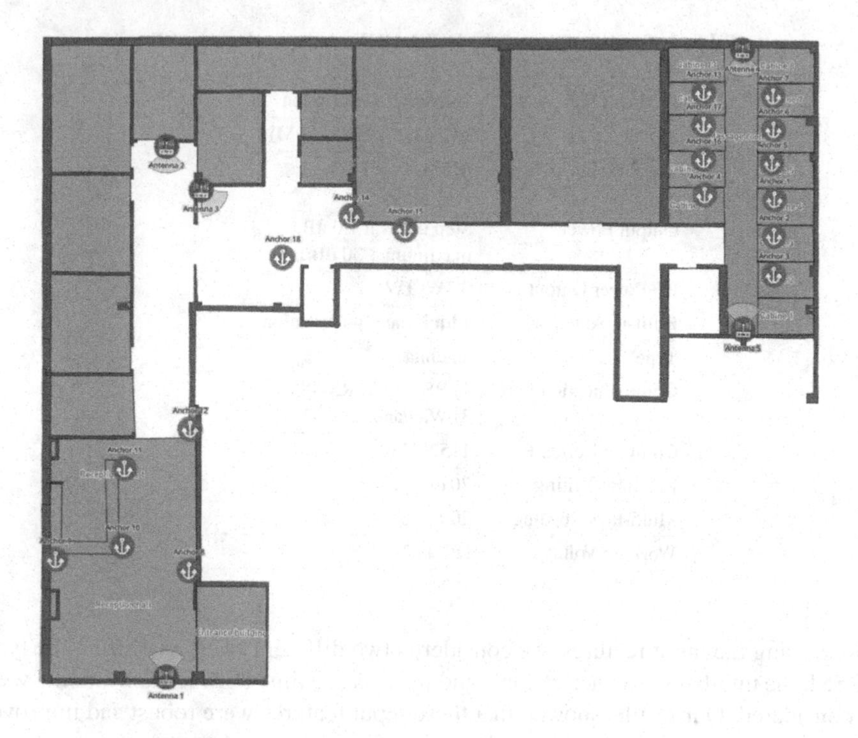

Fig. 2. The location of the antennas and anchor tags.

To achieve this analysis we collected data from all power levels for three days straight. We created a dataset with each single power level, and all the combinations of two power levels. Using those datasets we trained the best model from the previous work (Random Forest), and computed the Mean Absolute Error (MAE) for each dataset. Since this is a regression problem, we want to minimize the MAE, and at the same time maximize the number of tags detected. As such, the metrics score that we considered in this analysis is the subtraction the normalized MAE from normalized count of tags.

Figure 3 depicts the results of our analysis. In the diagonal, we can see the values of single power levels, while the remaining cells contains the metric value for each pair of power levels. It is possible to observe, that the pair $[210, 300]$mW offers the lowest error, while it captures the largest number of tags. Effectively, 300 mW will get all tags possible (max power), while 210 mW will provide distance discrimination.

The dataset that will be used to train the ML models are distributed in the following way (as depicted in Fig. 4), and had 68368 rows, each row representing a tag that was read by some antenna.

As shown in Fig. 4, the distribution of samples is not uniform for all distances, which is not the ideal, as it introduces a bias within the models. However, the major of the distances have a very good distribution (between 6.5% and 7.9%). To recap, we placed eighteen anchors that are able to produce twenty three different distances. This is possible as there are some overlaps between antennas, increasing the number of effective distances for a given set of anchor tags.

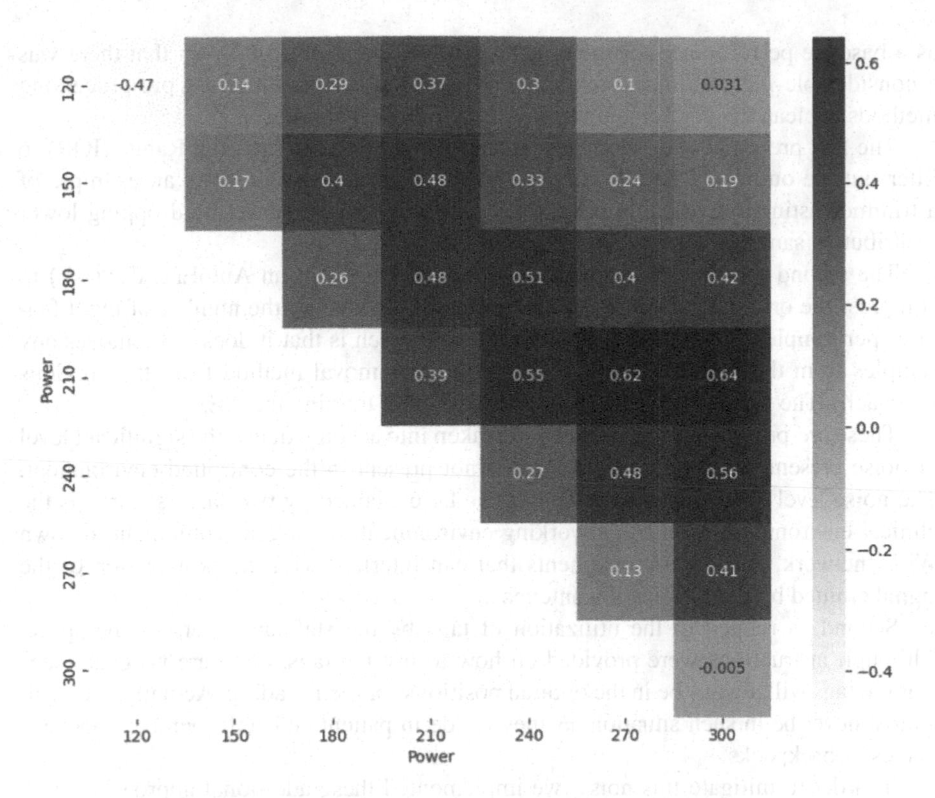

Fig. 3. Result of the power levels analysis impact on performance.

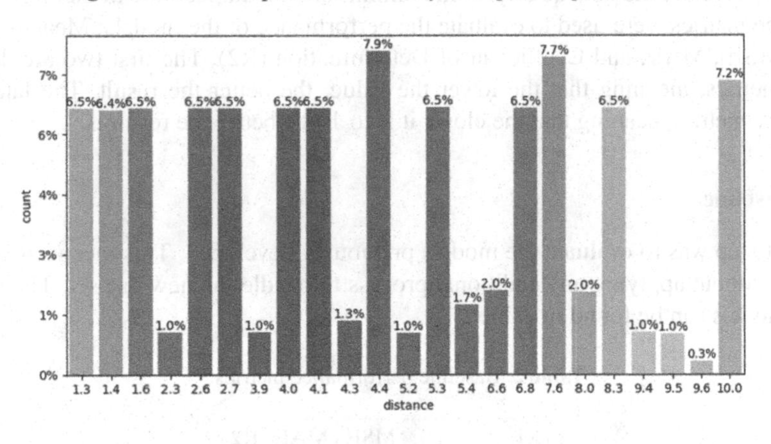

Fig. 4. Dataset Distribution.

6 Evolved Solution

We started by applying the ML models from the previous work, in the real-world scenario, as it showed capabilities without our requirements. These models would serve as a baseline performance for the system in a real environment. Given that there was a considerable increase in the models error, we applied two different pre-processing methods to clean the dataset and improve the model capabilities.

The first pre-processing method was simple applying Interquartile Range (IQR) to filter out the outliers. IQR is a measure of statistical dispersion, it is an example of a trimmed estimator, which improves the quality of the of dataset by dropping lower contribution samples.

The second pre-processing method implies the usage of an AutoEncoder (AE) to compress the original dataset into a smaller one, by reducing the number of input features per sample. The main advantage of this approach is that it does not removes any samples from the dataset (opposite to the outlier removal method from the previous approach). The main disadvantage is the necessity of training the AE.

These pre-processing approaches were taken into account due to the significant level of noise present in the dataset, which was not present in the controlled environment. The noise level in the real-world dataset can be explained by two factors. First, is the clinical environment itself. As a working environment, it contains equipment, its own Wi-Fi network, and other components that can interfere with the propagation of the signal emitted by the localization antennas.

Second, is related to the utilization of tags by the staff and users of the space. Although instructions were provided on how to use the tags, there are no guarantees that the tags will always be in the optimal positions for their reading. Actually, they will almost never be in such situation, as they reside in patient and staff members pockets, purses or backpacks.

In order to mitigate this noise, we implemented these additional approaches with the aim of refining the data quality, while minimizing the impact of external influences.

Three metrics were used to evaluate the performance of the models: Mean Squared Error (MSE), MAE, and Coefficient of Determination (R2). The first two are dissimilarity metrics, meaning that the lower the value, the better the result. The later is a similarity metric, meaning that the closer it is to 1, the better the result is.

6.1 Baseline

The first step was to evaluate the models previously developed. This means, using the models without applying any additional process to handle the new dataset. The results of the models can be found in Table 2.

Table 2. Baseline performance metrics.

Model	MSE	MAE	R2
LinearRegression	5.62	1.86	0.23
DecisionTree	1.64	0.57	0.78
RandomForest	2.23	1.03	0.70
KNeighbors	1.87	0.59	0.75
XGBoost	1.50	0.58	0.80
Voting Model	2.01	0.98	0.73

As presented in Table 2 the best possible value is 0.57 (MAE) achieved by the DT model. This represents a performance decrease compared to the best model from the previous work by an average of 0.02 m.

To better understand the model perform at different ranges, we computed the MAE for the best model (DT) at each possible distance and plotted them in the following histogram (see Fig. 5).

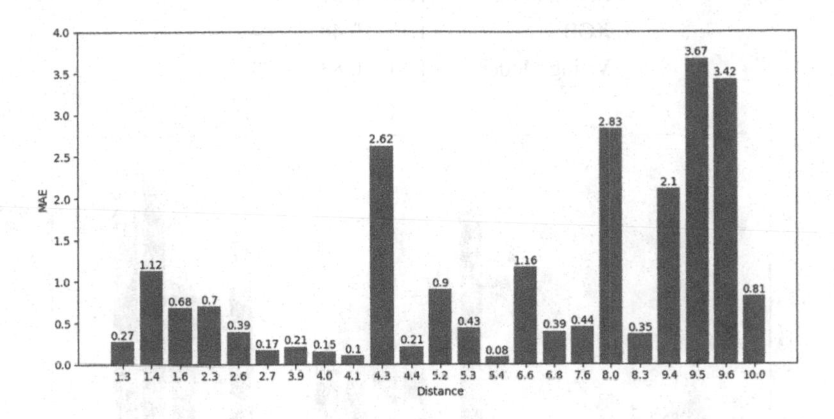

Fig. 5. MAE histogram at different ranges from the best baseline models (DT).

Through an analysis of the figures, we could observe that the model tends to make more errors in distances where there is less data, for distances greater than approximately 4 m. Comparing this with the distribution of the dataset (Fig. 4), it is quite noticeable.

6.2 First Pre-processing: IQR

The first approach to reduce the natural noise in the dataset was to apply the IQR outlier filtering method. This method calculates all the values between the first quartile (Q1) and the third quartile (Q3), giving more importance to the inner 50% of the data and eliminating the majority of outliers. The constant value used was 1.5, which is used to establish a threshold that determines how far the data can be considered "far" from the median. This value (1.5) is based on Tukey's fence, allowing for sensitive identification of potential outliers while remaining relatively conservative. The idea behind exploring the removal of outliers (data points with significantly different characteristics from what is considered normal) is to obtain a cleaner dataset.

With this approach pre-processing, we obtain the results presented in Table 3.

As we can see in Table 3 the best possible value is 0.47 (MAE) achieved by the DT model. This represents a performance increase compared to the best model from the initial training by an average of 0.1 m.

To better understand the model with the highest perform in this approach (DT) at different ranges, we computed the MAE for each possible distance and plotted them in the following histogram (see Fig. 6).

Table 3. IQR performance metrics.

Model	MSE	MAE	R2
LinearRegression	4.19	1.60	0.44
DecisionTree	1.27	0.47	0.83
RandomForest	1.88	0.94	0.75
KNeighbors	1.60	0.51	0.79
XGBoost	1.17	0.48	0.84
Voting Model	1.61	0.83	0.79

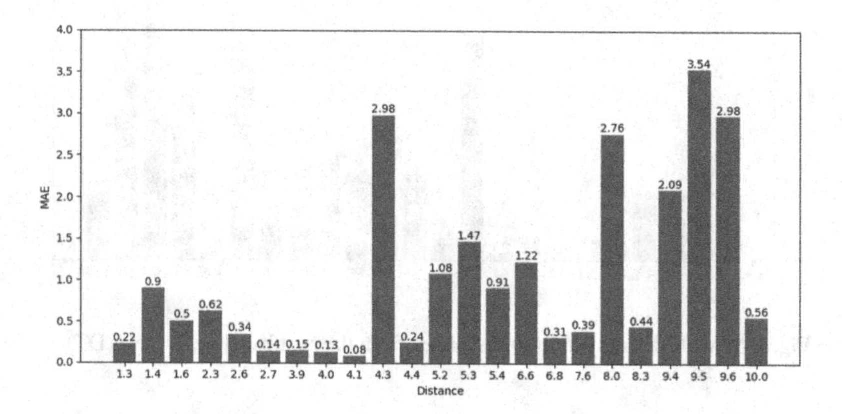

Fig. 6. MAE histogram at different ranges from the IQR best models (DT).

As in the previously approach (Baseline), it is evident that the model exhibits a higher error rate in distances with few data, particularly for distances exceeding approximately 4 m. This is aligned with the distribution of samples per range of the acquired dataset (as seen in Fig. 4).

6.3 Second Pre-processing: AutoEncoder

We also implemented an AutoEncoder (AE) [13] for feature reduction. An AutoEncoder is a neural network that compresses the input data into a lower-dimensional representation, and then reconstructs it from that lower dimension.

This network is composed of two different blocks, the encoder and the decoder. The encoder is a initial part of the network that compresses the input into a latent space with a smaller dimension. The decoder tries to reconstruct the input signal from the constrained latent space. The network is trained on the input data, while trying to reproduce it using a constrained latent space. The loss function is the difference between the input features and the reconstructed features. After an initial training phase, we only use the encoder part of the network to compress the input features into a lower-dimensional called latent space. One consequence of compressing the input features into a lower-dimensional latent space is the reduction of noise in the input signals.

After training the AutoEncoder, we simple use the encoder part of the network to transform the dataset into a new one on the constrained latent space. As previously mentioned, with reduced noise levels. After we just apply the ML models naturally.

With this pre-processing method, we obtained the results presented in Table 4.

Table 4. AutoEncoder performance metrics.

Model	MSE	MAE	R2
LinearRegression	5.70	1.93	0.22
DecisionTree	2.71	1.02	0.64
RandomForest	3.15	1.27	0.57
KNeighbors	2.05	0.64	0.72
XGBoost	2.05	0.84	0.72
Voting Model	2.49	1.14	0.66

As we can see in Table 4 the best possible value is 0.64 (MAE) achieved by the KNearest Neighbors (K-NN) model. This represents a performance decrease compared to the best model from the initial training by an average of 0.07 m, and a performance decrease compared to the results in the IQR approach by 0.17 m.

In order to better understand how the best model performs (K-NN model) in the existing ranges, we plotted it (see Fig. 7), similarly to the previous approaches.

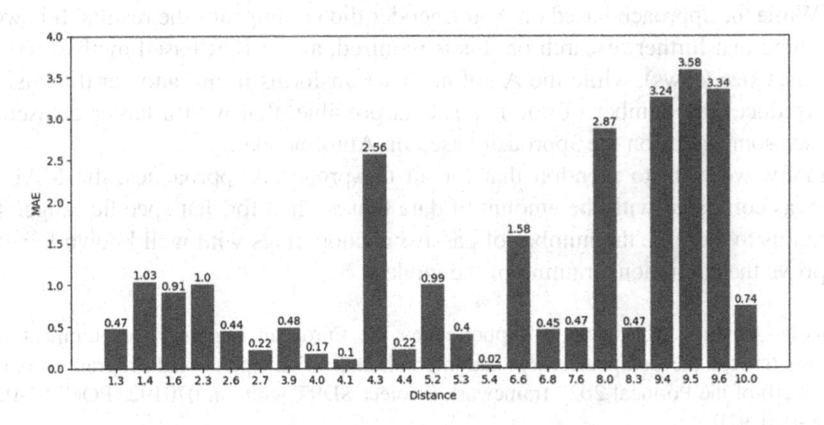

Fig. 7. MAE histogram at different ranges from the AE best model (K-NN).

Similarly to the previous two approaches (baseline and IQR) it is apparent that the model demonstrates increased error rates in ranges where training data is limited. Again, especially visible for distances greater than approximately 4 m, and aligned with the distribution of samples per range of the acquired datasets.

7 Conclusion

The management of health systems has been one of the main challenges in several European countries, especially where the ageing population is increasing. One of the technologies adopted is active location solutions, which allows the staff within the hospital to quickly find any sort of entity, from key persons to equipment.

In this work, we evaluated the usage of dedicated hardware (namely the RF antenna) for indoor location within a medical environment. From a previous work we devise two new feature (number of activation's and the average time of activation's for passive tags) that in conjunction with RSSI produce models that are more robust. We also established a performance based, based on data acquired on a controlled environment.

The current work is aligned with collaboration with medical entities, and as such we were able to acquired data in a real-world health clinic.

In this work, we present the details of the scenario, and how we selected the ideal power levels to explore with the antennas. It is a compromise between the amount of data acquired from the antennas and the latency of the overall system.

Following that, we evaluated the previous solution to compare the current results with the previous ones. Simply applying the previous models lead to a decrease of 0.02 m in performance, when taking into account the best ML models.

Our analysis show an increase noise level within the dataset. That noise is a consequence of the environment itself, arising from moving people, sub-optimal tag location and orientation and othre interference. To deal with this we applied two different pre-processing approaches: outlier removal with IQR and noise reduction with AutoEncoder.

The approach based on IQR was able to improve the results, achiving an MAE of 0.47. While the approach based on AutoEncoder did not improve the results. It is worth mentioning that further research on this is required, as the IQR based method reduces the dataset size (rows), while the AutoEncoder transforms it into another dimensional space (reduces the number of columns). Is is possible, that with a larger dataset, we could see some gains on the approach based on AutoEncoder.

Finally, we want to mention that for all the proposed approaches, the MAE per range was correlated with the amount of data that we had for that specific range. This can lead us to increase the number of passive anchors (tags with well known location) to improve the continuous training of the models.

Acknowledgements. This work is supported by the European Regional Development Fund (FEDER), through the Competitiveness and Internationalization Operational Programme (COMPETE 2020) of the Portugal 2020 framework [Project SDRT with Nr. 070192 (POCI-01-0247-FEDER-070192)]

References

1. Carotenuto, R., Merenda, M., Iero, D., Della Corte, F.G.: An indoor ultrasonic system for autonomous 3-d positioning. IEEE Trans. Instrum. Meas. **68**(7), 2507–2518 (2019)
2. Chanama, L., Wongwirat, O.: A comparison of decision tree based techniques for indoor positioning system. In: 2018 International Conference on Information Networking (ICOIN), pp. 732–737 (2018)

3. Choi, J.S., Lee, H., Elmasri, R., Engels, D.W.: Localization systems using passive UHF RFID. In: 2009 Fifth International Joint Conference on INC, IMS and IDC, pp. 1727–1732 (2009)
4. Hightower, J., Vakili, C., Borriello, G., Want, R.: Design and calibration of the spoton ad-hoc location sensing system. unpublished, 31 August 2001
5. Ni, L., Liu, Y., Lau, Y.C., Patil, A.: Landmarc: indoor location sensing using active RFID. In: Proceedings of the First IEEE International Conference on Pervasive Computing and Communications, 2003. (PerCom 2003), pp. 407–415 (2003)
6. Paiva, S., Brito, D., Leiva-Marcon, L.: Real time location systems adoption in hospitals-a review and a case study for locating assets. Acta Sci. Med. Sci. **2**(7), 02–17 (2018)
7. Pomárico-Franquiz, J., Khan, S.H., Shmaliy, Y.S.: Combined extended FIR/Kalman filtering for indoor robot localization via triangulation. Measurement **50**, 236–243 (2014)
8. Santos., R., et al.: Towards improved indoor location with unmodified RFID systems. In: Proceedings of the 12th International Conference on Pattern Recognition Applications and Methods - ICPRAM, pp. 156–163. INSTICC, SciTePress (2023)
9. Shen, H., Ding, Z., Dasgupta, S., Zhao, C.: Multiple source localization in wireless sensor networks based on time of arrival measurement. IEEE Trans. Signal Process. **62**(8), 1938–1949 (2014)
10. Suroso, D.J., Rudianto, A.S., Arifin, M., Hawibowo, S.: Random forest and interpolation techniques for fingerprint-based indoor positioning system in un-ideal environment. Int. J. Comput. Digit. Syst. (2021)
11. Tegou, T., Kalamaras, I., Votis, K., Tzovaras, D.: A low-cost room-level indoor localization system with easy setup for medical applications. In: 2018 11th IFIP Wireless and Mobile Networking Conference (WMNC), pp. 1–7 (2018)
12. Vorst, P., Schneegans, S., Yang, B., Zell, A.: Self-localization with RFID snapshots in densely tagged environments. In: 2008 IEEE/RSJ International Conference on Intelligent Robots and Systems, pp. 1353–1358 (2008)
13. Wang, W., Huang, Y., Wang, Y., Wang, L.: Generalized autoencoder: a neural network framework for dimensionality reduction. In: Proceedings of the IEEE Conference on Computer Vision and Pattern Recognition (CVPR) Workshops, June 2014
14. Wilson, P., Prashanth, D., Aghajan, H.: Utilizing RFID signaling scheme for localization of stationary objects and speed estimation of mobile objects. In: 2007 IEEE International Conference on RFID, pp. 94–99 (2007)
15. Xiong, J., Jamieson, K.: ArrayTrack: a fine-grained indoor location system. In: 10th USENIX Symposium on Networked Systems Design and Implementation (NSDI 13), pp. 71–84. USENIX Association, Lombard, IL, April 2013

Applications

MSAA-Net: Multi-Scale Attention Assembler Network Based on Multiple Instance Learning for Pathological Image Analysis

Takeshi Yoshida[1], Kazuki Uehara[2,3](✉), Hidenori Sakanashi[1,3], Hirokazu Nosato[3], and Masahiro Murakawa[1,3]

[1] University of Tsukuba, 1-1-1, Tennoudai, Tsukuba, Ibaraki, Japan
yoshida.takeshi.sp@alumni.tsukuba.ac.jp
[2] University of the Ryukyus, 1, Senbaru, Nakagami, Nishihara, Okinawa, Japan
k-uehara@grs.u-ryukyu.ac.jp
[3] National Institute of Advanced Industrial Science and Technology (AIST), 1-1-1, Umezono, Tsukuba, Ibaraki, Japan
{h.sakanashi,h.nosato,m.murakawa}@aist.go.jp

Abstract. In this paper, we present a multi-scale attention assembler network (MSAA-Net) tailored for multi-scale pathological image analysis. The proposed approach identifies essential features by examining pathological images across different resolutions (scales) and adaptively determines which scales and spatial regions predominantly influence the classification. Specifically, our approach incorporates a two-stage feature integration strategy. Initially, the network allocates the attention scores to relevant local regions of each scale and then refines the attention scores for each scale as a whole. To facilitate the training of the MSAA-Net, we employ the technique of multiple instance learning (MIL), which enables us to train the classification model using the pathologist's daily diagnoses of whole slide images without requiring detailed annotation (i.e., pixel-level labels), thereby minimizing annotation effort. We evaluate the effectiveness of the proposed method by conducting classification experiments using two distinct sets of pathological image data. We conduct a comparative analysis of the attention maps generated by these methods. The results demonstrate that the proposed method outperforms state-of-the-art multiscale methods, confirming the effectiveness of MSAA-Net in classifying multi-scale pathological images.

Keywords: Multi-scale whole slide image · Multiple instance learning · Pathological image analysis · Computer-aided diagnosis

1 Introduction

A pathological assessment is essential in cancer medical care as it determines the treatment direction. Pathologists analyze a sample by changing magnification levels on a microscope and make a diagnosis based on histopathological aspects, such as cell size

T. Yoshida, K. Uehara—These authors contributed equally to this manuscript.

M. De Marsico et al. (Eds.): ICPRAM 2023, LNCS 14547, pp. 49–68, 2024.
https://doi.org/10.1007/978-3-031-54726-3_4

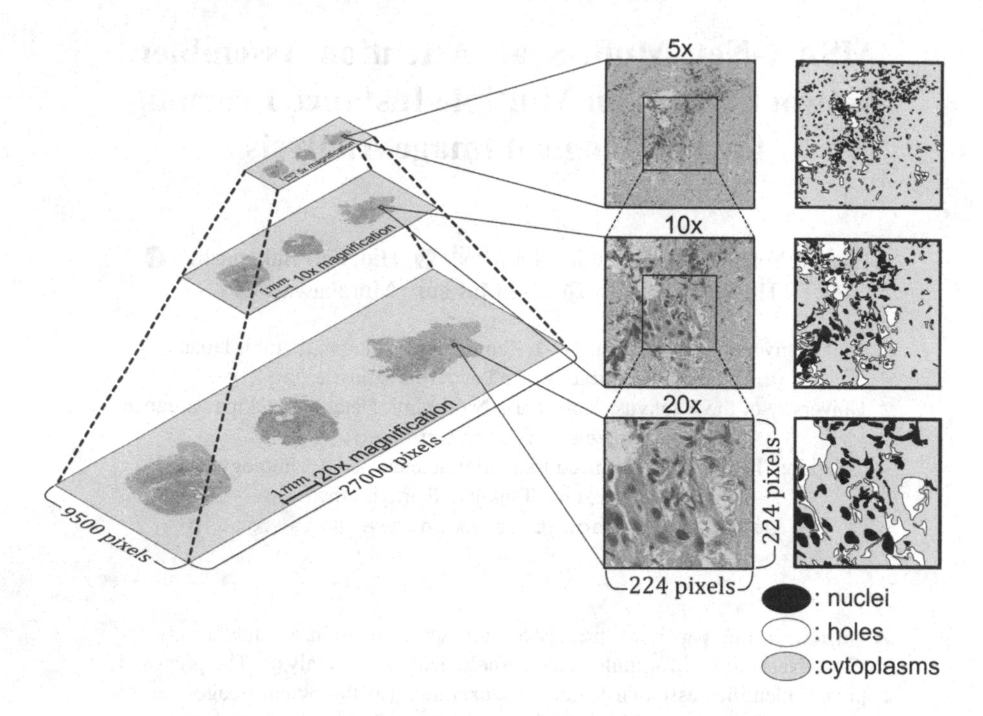

Fig. 1. Illustration of the specimen at different magnifications.

and morphology, nuclei, and tissue arrangement. Lately, examining the whole slide images (WSIs), depicted in Fig. 1, has taken over the traditional microscope-based observation. The WSIs are digital pathology images captured digitally by scanning the entire glass slide under high magnification. By downsampling the high-magnification image to a low-magnification image, WSIs allow observations in various scales similar to traditional microscopes. Although diagnostic support technology has advanced, the workload on pathologists remains high due to the limited number of professionals in the field.

In this scenario, developing an automated pathological diagnostic support system using machine learning techniques is explored [4,5,17]. In such research, classification approaches are utilized to determine whether each WSI has potential cancer cells, and these strategies employ multiple instance learning (MIL) [6,15]. MIL is an approach that divides one large image (e.g., WSI) into many small image patches and trains a machine-learning model from a single label (e.g., presence or absence of cancer) given to these entire patches. The advantage of MIL is that by dividing a high-resolution WSI into small image patches, it is possible to analyze the image in detail without requiring a large amount of computational resources. Furthermore, the annotation effort is greatly reduced because only the pathologist's diagnosis (e.g., WSI-level labels) for the entire WSI is used, rather than requiring labels for each image patch.

The optimal magnification for diagnosis depends on the type of histopathological feature. Therefore, a multi-scale approach is expected to improve the accuracy of WSI classification. For instance, examining at the cellular level, such as the condition of nuclei, and at the tissue level, such as the tissue structure consisting of the arrangement of cells, are appropriate at high and low magnification, respectively. The benign and cancerous tissues' examples at different magnifications are shown in Fig. 2. Figure 2a shows the large nuclei as the abnormal cellular characteristics at the 20x magnification, while Fig. 2b shows the alignment of the large nuclei around the holes as the abnormal structural characteristics at the 10x magnification.

Consequently, we propose a multi-scale attention assembler network (MSAA-Net) designed to focus on crucial areas at each scale and emphasize the appropriate scale for classification. To integrate the features at the patch level of the image with the features at the WSI level and to leverage the multi-scale approach, we adopt a two-step feature aggregation method that includes a region-wise aggregation at each scale and an overall scale integration. First, based on the attention scores, the region aggregator derives region-level features for each scale. High coefficients are assigned when the corresponding regions are important for each scale. Next, the scale aggregator merges the scale-level features with the WSI-level feature. It relies on the attention weights of an essential scale to perform the classification.

We experimentally evaluated the proposed method on two datasets obtained from a public database and clinical cases. The results indicated that MSAA-Net achieved better classification than the conventional methods using the single and multi-scale WSIs in both datasets. Notably, our approach improved the accuracy of cancer detection by approximately 20% compared to the conventional methods.

This paper explains the details of MSAA-Net, our previous work [20], and further experiments and analysis. The remainder of this paper is organized as follows: Sect. 2 provides a literature review on MIL in pathological image analysis. Section 3 presents the proposed method, MSAA-Net. Section 4 describes the experiments comparing its performance with the conventional methods. Section 5 discusses the results and attention maps generated by the proposed and conventional methods. Finally, Sect. 6 concludes the paper by summarizing the key findings.

2 Related Work

2.1 WSI Classification

WSI classification assigns labels to WSIs that indicate the type of cancer, the presence or absence of cancer, and so on. One of these methods is to classify each image patch divided from WSIs and integrate the results to classify WSI [3, 14]. These methods offer detailed diagnostic results for each image patch, similar to semantic segmentation. However, they require fine-grained annotation, namely the diagnosis of many image patches. Therefore, the cost of preparing training data is enormous. Another approach has been proposed to classify WSI compressed to a computationally feasible size [19]. Although detailed annotation at the patch level is not required, small anomalies can be missed because of the drop in detail due to the compression of the image.

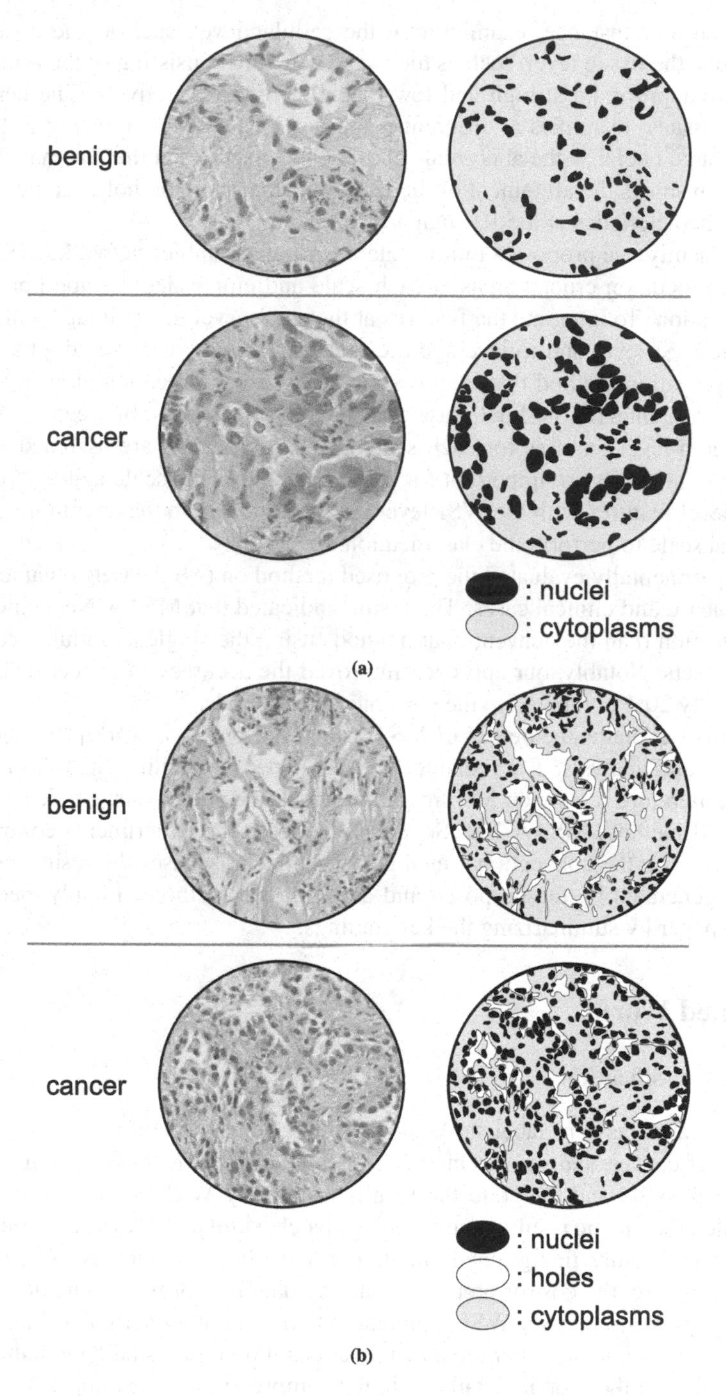

Fig. 2. Examples of pathological tissue image. (a) Comparison of benign and cancerous tissue at the 20x magnification. (b) Comparison of benign and cancerous tissue at the 10x magnification.

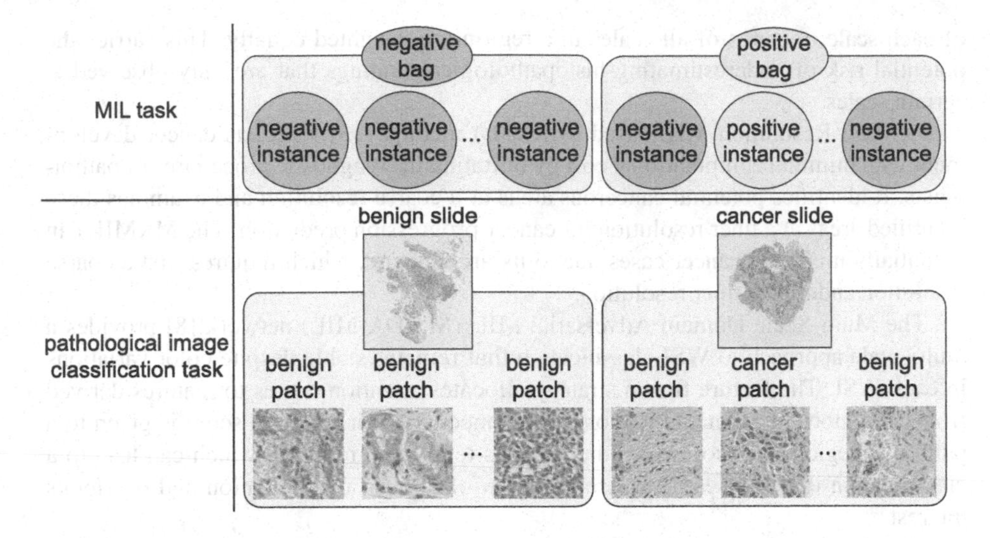

Fig. 3. Correspondence of the input data for the MIL and WSIs classification tasks.

2.2 MIL-Based WSI Classification

Single-Scale Classification. MIL trains a model that solves the task of predicting the class of an entire group, called a bag, consisting of multiple instances. Classifying WSIs can be considered a MIL problem, and many studies have been conducted [2,7,9,16]. Specifically, instances are image patches divided from WSIs; bags are the WSIs, and bag labels are diagnostic labels (Fig. 3). The methods generally calculate instance features and summarize them to bag-level features by feature aggregation mechanisms such as mean pooling and maximum pooling. Then, they predict bag-level labels based on the aggregated features. However, this aggregation mechanism may ignore some useful instance features because they treat all features uniformly.

To solve this problem, Attention-based Deep MIL (ADMIL) [10] has been proposed, which uses an attention mechanism [21] for feature aggregation. The attention mechanism determines an attention weight for each instance feature and calculates bag-level features utilizing a weighted sum. As a result, a higher attention weight indicates a higher contribution of the bag-level feature, i.e., it is more important for the classification. However, although utilizing the appropriate scale for more accurate pathological diagnosis is important, this method cannot handle the multi-scale nature of WSIs.

Multi-scale Classification. MIL-based methods, which consider the multiscale nature of WSI, have been proposed [8,12,13]. The Dual Stream MIL Network (DSMIL) [12] classifies multi-scale WSIs by estimating regions of interest based on histopathological appearance. DSMIL concatenates the feature vectors of all scales corresponding to each region and calculates the attention value for each concatenated vector based on the feature space distance. Because multi-scale WSIs are not observed from the perspective of each scale, features of all scales in a region are weighted equally. This carries the potential risk of underestimating histopathological findings that are only observed at certain scales.

A Multi-Resolution MIL-based (MRMIL) method [13] forecasts cancer development with minimal computational cost by imitating the diagnostic procedures of pathologists. It identifies potential cancerous areas at a coarse resolution and examines these identified areas at a finer resolution for cancer progression prediction. The MRMIL can potentially miss tiny cancer cases due to its architecture, which requires first a coarse resolution and then a finer resolution.

The Multi-Scale Domain Adversarial MIL (MS-DA-MIL) network [8] provides a multi-scale approach to WSI classification that remains stable despite color variations in each WSI. The feature fusion strategy allocates attention scores to features derived from all regions at all magnifications simultaneously. When a high score is given to a particular region, the scores for other regions tend to be reduced, which can lead to a classification that focuses on a certain region of a given magnification and overlooks the rest.

3 Proposed Method

To overcome the limitations of the conventional methods, we adopt a two-stage feature aggregation approach, which evaluates the crucial areas within the tissue of each magnification image and then determines the important scale among each magnification image. In the first step, an image at each magnification is observed independently to identify areas crucial for classification, considering the relationships between tissue structures and between cells. It can, therefore, identify smaller cancers seen on high-magnification images. In the subsequent step, the areas important for classification within the tissue of each magnification image are compared to each other to select the appropriate scale for the target. This means that even without knowing which classification target is contained in the WSIs, the method can determine the critical magnification from the input images.

3.1 Problem Formulation

A given WSI $\boldsymbol{X}_i (i = 1, \ldots, N)$ composed of multiple images at scale $s_j (j = 1, \ldots, S)$ is segmented into $n_i^{(s_j)}$ image patches $\boldsymbol{x}_{ik}^{(s_j)} (k = 1, \ldots, n_i^{(s_j)}) \in \boldsymbol{X}_i$, each having the size $W \times H$, where W and H represent width and height respectively. The label of each image patch at each scale is expressed using a one-hot encoding as follows:

$$y_{ikl}^{(s_j)} = \begin{cases} 1 & \text{if } l = c \\ 0 & \text{otherwise} \end{cases} \qquad (1)$$
$$(l = 1, \ldots, C),$$

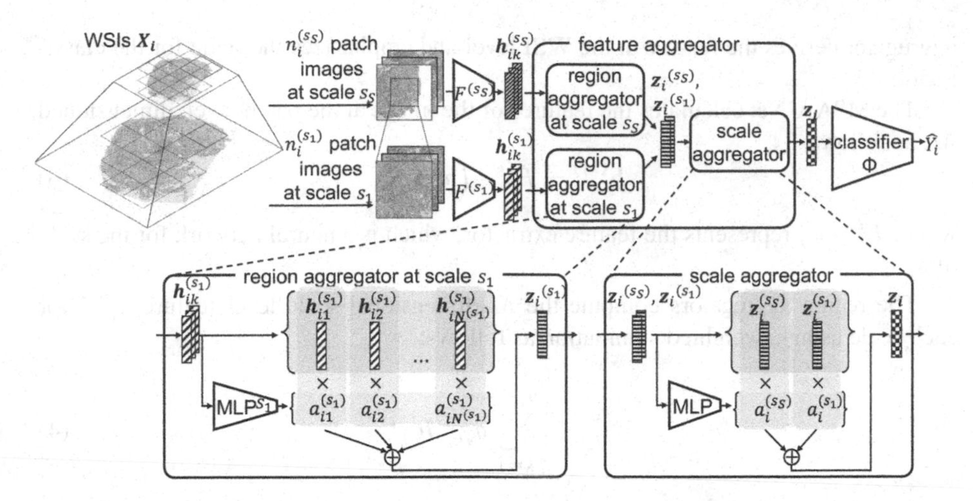

Fig. 4. Illustration of the architecture of the proposed MSAA-Net. This figure is modified based on the paper [20].

where c and C are the index of the class and the number of classes, respectively. The image patch containing the cancerous area is assigned to the cancer class. The WSI level label Y_{il}, which is the one-hot encoding, is then defined as follows:

$$Y_{il} = \begin{cases} 1 & \text{if } \sum_{j=1}^{S} \sum_{k=1}^{n_i^{(s_j)}} y_{ikl}^{(s_j)} > 0 \\ 0 & \text{otherwise} \end{cases} \quad (l = 1, \ldots, C). \tag{2}$$

During the training process, only the WSI-level label Y_{il} is available, and the learner cannot access the label for image-patch y_{ikl}.

3.2 Multi-scale Attention Assembler Network

Figure 4 illustrates the architecture of the MSAA-Net. The network comprises feature extractors $F^{(s_j)}(\cdot)$ that are tailored for each scale s_j, a feature aggregator, and a classifier $\Phi(\cdot)$. The feature aggregator consists of region aggregators for each scale and scale aggregator. The proposed network determines the labels by sequentially processing the target WSIs through feature extraction, aggregation, and classification processes. During the feature aggregation phase, we employ a two-stage feature aggregation strategy, sequentially using scale-specific region aggregators and then the scale aggregator to realize multi-scale analysis approaches. At first, each scale's independent region aggregators determine the attention weights that correspond to the features derived from their respective scale perspectives. Using a weighted sum, these region aggregators identify region-level features for their respective scales, highlighting the key features. Subsequently, the scale aggregator adaptively determines the attention weights for the region-level features of each scale based on the specific classification objective. The scale aggregator derives the feature at the WSI level and emphasizes the scale for the classification.

The MSAA-Net calculates the features of the image at the patch level, dimensioned M, as follows:

$$\boldsymbol{h}_{ik}^{(s_j)} = F^{(s_j)}\left(\boldsymbol{x}_{ik}^{(s_j)}\right), \tag{3}$$

where $F^{(s_j)}(\cdot)$ represents the feature extractor, which is a neural network for the scale of s_j.

The region aggregators compute the M-dimensional scale-level feature $\boldsymbol{z}_i^{(s_j)}$ for each scale using a weighted summation as follows:

$$\boldsymbol{z}_i^{(s_j)} = \sum_{k=1}^{n_i^{(s_j)}} a_{ik}^{(s_j)} \boldsymbol{h}_{ik}^{(s_j)}, \tag{4}$$

where $a_{ik}^{(s_j)}$ represents the attention weight, with a higher value indicating greater importance of the associated feature $\boldsymbol{h}_{ik}^{(s_j)}$ in the classification. The attention weight $a_{ik}^{(s_j)}$ is derived from the features using the Multi-Layer Perceptron (MLP) combined with a softmax function for each scale as follows:

$$a_{ik}^{(s_j)} = \frac{\exp\{\boldsymbol{w}^{(s_j)\mathsf{T}} \tanh\left(\boldsymbol{V}^{(s_j)} \boldsymbol{h}_{ik}^{(s_j)\mathsf{T}}\right)\}}{\sum_{l=1}^{n_i^{(s_j)}} \exp\{\boldsymbol{w}^{(s_j)\mathsf{T}} \tanh\left(\boldsymbol{V}^{(s_j)} \boldsymbol{h}_{il}^{(s_j)\mathsf{T}}\right)\}}, \tag{5}$$

where $\boldsymbol{w}^{(s_j)}$ and $\boldsymbol{v}^{(s_j)}$ represent the $L \times 1$ and $L \times M$ dimensionally trainable parameters of the MLP for each specific scale, respectively.

The scale aggregator computes the WSI-level feature $\boldsymbol{z}_i$ by employing a weighted sum, similar to the method used by the region aggregator as follows:

$$\boldsymbol{z}_i = \sum_{k=1}^{S} a_i^{(s_k)} \boldsymbol{z}_i^{(s_k)}. \tag{6}$$

The attention weight $a_i^{(s_k)}$ represents the contribution of each scale to the WSI-level feature. Specifically, the attention weight $a_i^{(s_k)}$ is determined by the MLP in the manner as follows:

$$a_i^{(s_k)} = \frac{\exp\{\boldsymbol{w}^{\mathsf{T}} \tanh\left(\boldsymbol{V} \boldsymbol{z}_i^{(s_k)\mathsf{T}}\right)\}}{\sum_{j=1}^{S} \exp\{\boldsymbol{w}^{\mathsf{T}} \tanh\left(\boldsymbol{V} \boldsymbol{z}_i^{(s_j)\mathsf{T}}\right)\}}, \tag{7}$$

where $\boldsymbol{w}$ and $\boldsymbol{V}$ represent trainable parameters of the MLP with dimensions $L \times 1$ and $L \times M$, respectively. Their values differ from those in Eq. 5.

The label probabilities, represented as $\hat{Y}_i = \left\{\hat{Y}_{i1}, \ldots, \hat{Y}_{iC}\right\}$, are determined by the linear classifier $\Phi(\cdot)$ as follows:

$$\begin{aligned} \hat{Y}_i &= \Phi\left(\boldsymbol{z}_i\right) \\ &= \boldsymbol{w}\boldsymbol{z}_i + \boldsymbol{b}, \end{aligned} \tag{8}$$

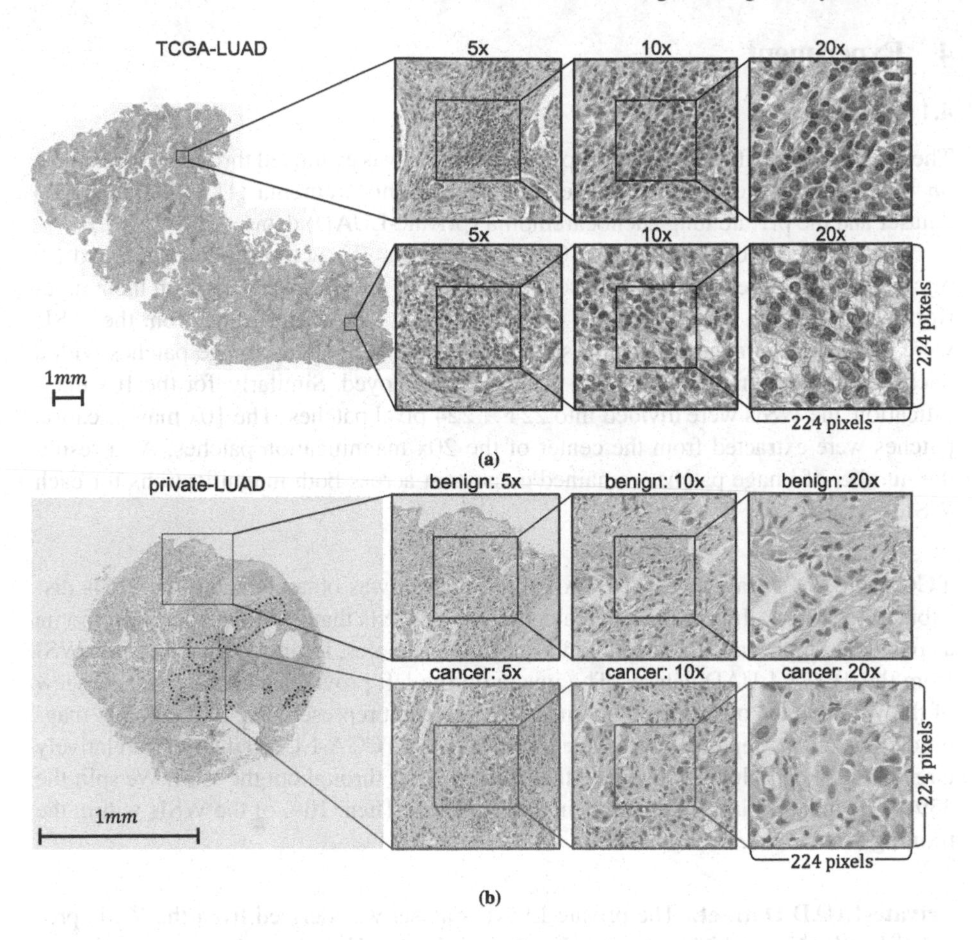

Fig. 5. (a) WSI included in TCGA-LUAD dataset. (b) WSI included in the private-LUAD dataset. This figure is from the paper [20].

where $\boldsymbol{w}$ and $\boldsymbol{b}$ indicating weight and bias, respectively.

The optimization of the MSAA-Net during training is achieved by minimization of the cross-entropy loss as follows:

$$\mathcal{L} = -\frac{1}{N}\sum_{i=1}^{N}\sum_{l=1}^{C} Y_{il}\log \hat{Y}_{il}. \tag{9}$$

These equations are adapted from Yoshida et al. [20].

4 Experiment

4.1 Dataset

The classification efficacy of our proposed approach was evaluated through experiments on two datasets: the cancer genome atlas lung adenocarcinoma [1] (TCGA-LUAD) dataset and the private lung adenocarcinoma (private-LUAD) dataset.

Both datasets included 20x and 10x magnifications. This corresponds to 0.5 and 1.0 microns per pixel, respectively. These magnifications were chosen based on their superior performance in the preliminary experiment. For the 20x magnification, the WSIs were divided into image patches of size 224 × 224 pixels. Only image patches with a background region ratio of less than 40% were employed. Similarly, for the 10x magnification, the WSIs were divided into 224 × 224 pixel patches. The 10x magnification patches were extracted from the center of the 20x magnification patches. As a result, the number of image patches remained consistent across both magnifications for each WSI.

TCGA-LUAD Dataset. The TCGA-LUAD dataset was obtained from the WSIs distributed by TCGA. In addition, we categorized the WSIs that had lung adenocarcinoma as positive samples and those without as negative samples. Figure 5 (a) depicts the WSI from the TCGA-LUAD dataset. The image on the left provides a comprehensive view of the WSI. On the other hand, the images on the right represent 5x, 10x, and 20x magnifications of two separate areas. The WSI from the TCGA-LUAD dataset is relatively large, and the pathologist confirmed that cancer exists throughout the WSI. We split the 359 WSIs into training and test sets in a 65:35 ratio. Then, 10% of the WSIs within the training set were used for validation.

Private-LUAD Dataset. The private LUAD dataset was derived from the WSIs provided by the Nagasaki University Hospital in Japan. This dataset was a curation of lung biopsy specimens. As a result, the WSIs in this dataset contain smaller regions of cancer, which makes their classification more complicated than in the TCGA-LUAD dataset. Similarly, we categorized WSIs with lung adenocarcinoma as positive samples and those without as negative samples, mirroring the approach used for the TCGA-LUAD dataset. Figure 5 (b) illustrates the WSI from the private-LUAD dataset. The image on the left provides the overall view of the WSI. Unlike the TCGA LUAD dataset, the private-LUAD dataset is provided with annotations at the pixel level by pathologists for analytical purposes (not used during training).

Areas with dotted outlines were diagnosed as adenocarcinoma. In contrast, the figures on the right show 5x, 10x, and 20x magnifications of two different diagnostic areas. The WSIs in the private-LUAD dataset are compact, with smaller areas of cancer. We partitioned the 863 WSIs into the training and test subsets with 90:10 ratios. Then, 10% of the WSIs from the training subset were assigned for validation. To deal with data skewness, we increased the number of positive WSIs in the training subset using data augmentation techniques. For this augmentation, tissue sections were randomly rotated and overlaid on a blank background. Table 1 provides the number of WSIs for each subset in both experimental datasets.

Table 1. Assignment of the WSIs in both datasets for the experiments. This table adapted from Yoshida et al. [20].

Dataset	Training set			Validation set			Test set		
	All	Negative	Positive	All	Negative	Positive	All	Negative	Positive
TCGA-LUAD	208	122	86	27	16	11	124	37	87
private-LUAD	717	359	358	80	39	41	93	45	48

4.2 Comparison Methods and Evaluation Metrices

Two different experiments were conducted. First, we evaluated the classification accuracy of our proposed method against single-scale methods to validate the advantages of the multi-scale strategy. Both the single-scale method (DA-MIL network [8]) and our method used the same backbone network. We then verified the usefulness of the two-stage feature aggregation mechanism within the MSAA-Net. We compared the classification accuracy of the DSMIL, the MS-DA-MIL network, and the MSAA-Net. The DA-MIL network served as the feature extractor for these networks, consistent with the first experiment.

As evaluation metrics, we used precision, recall, and F1 score, and they are calculated as follows:

$$\text{Precision} = \frac{TP}{TP + FP}, \tag{10}$$

$$\text{Recall} = \frac{TP}{TP + FN}, \tag{11}$$

$$\text{F1 score} = \frac{2 \times \text{Precision} \times \text{Recall}}{\text{Precision} + \text{Recall}}, \tag{12}$$

where TP is the number of true positives, FP is the number of false positives, FN is the number of false negatives, and TN is the number of true negatives. True positive refers to the number of positive WSIs that are accurately identified as positive by the classification method. Subsequently, the false positive refers to the number of negative WSIs falsely identified as positive, while the false negative refers to the positive WSIs falsely classified as negative by the classification method. Finally, the true negative represents the negative WSIs that are correctly classified as negative. Precision is the fraction of the retrieved items that are relevant, whereas recall is the fraction of the relevant items that are retrieved. The F1 score is the weighted harmonic mean of precision and recall. Therefore, a higher F1 score is associated with a lower rate of missed and overdetected cancers. In the second experiment, we compared the classification accuracy by the average and standard deviation of each evaluation metric by the five-fold cross-validation. These experiments were conducted using five different sets of training and validation data to ensure that WSIs were not repeated in any single trial.

4.3 Implementation Details

To compare with the single-scale methods, we created a dataset for training the DA-MIL network by extracting 100 image patches from tissue regions in WSIs, similar

Table 2. Results of the conventional single scale DA-MIL network in comparison to the proposed MSAA-Net on two datasets.

Dataset	Method	magnifications	TP	FP	FN	TN
TCGA-LUAD	DA-MIL	20x	85	4	2	33
		10x	84	2	3	35
	MSAA-Net(ours)	20x-10x	83	1	4	36
private-LUAD	DA-MIL	20x	45	27	3	18
		10x	36	1	12	44
	MSAA-Net(ours)	20x-10x	43	1	5	44

Table 3. Comparison of the summarized results of the single-scale DA-MIL network and the MSAA-Net on two datasets. This table is from the paper [20].

Dataset	Method	magnifications	F1	Precision	Recall
TCGA-LUAD	DA-MIL	20x	0.966	0.955	**0.977**
		10x	**0.971**	0.977	0.966
	MSAA-Net(ours)	20x-10x	**0.971**	**0.988**	0.954
private-LUAD	DA-MIL	20x	0.750	0.625	**0.938**
		10x	0.847	0.973	0.750
	MSAA-Net(ours)	20x-10x	**0.935**	**0.977**	0.896

to the paper [8]. Additionally, we used VGG16 [18] feature extractor with two fully connected layers for the network [8]. We then used the trained DA-MIL (VGG16) as the feature extraction component for the MSAA-Net, DSMIL, and MS-DA-MIL networks.

In a comparative experiment using multi-scale techniques, we used the model architecture of the MS-DA-MIL network, mirroring the configuration in the paper [8]. We then retained the original model framework of the DSMIL [12], except for the feature extractor for the network. In MSAA-Net, we adopted the identical network architecture of the region aggregator over each scale as well as the scale aggregator. Those aggregators included a sequence of the linear layer, hyperbolic tangent activation, and another linear layer, followed by the softmax function, which determined the attention weights. Finally, we employed the single-layer perceptron for classification.

All models were trained using automatic mixed precision, gradient accumulation, and the Adam optimizer as described in [11]. We set a mini-batch size of 16. For the comparison study, we set the training epochs to 50 for single-scale and 100 for multi-scale methods.

4.4 Results

Table 2 and Table 3 display the classification results for the single-scale method and our proposed method. Compared to the conventional method, our method was either equal to or better than the conventional method for all metrics in both datasets. Specifically, our proposed approach showed an 18.5% improvement in the F1 score compared to the DA-MIL network at 20x magnification within the private-LUAD dataset. The method significantly reduced the number of false positives and maintained a minimal number

Table 4. Comparison of classification results using conventional multiscale methods and the proposed MSAA-Net on two datasets.

Dataset	Method	TP	FP	FN	TN
TCGA-LUAD	DSMIL	84.6	2.4	2.4	34.6
	MS-DA-MIL	84	2.6	3	34.4
	MSAA-Net(ours)	84	2.2	3	34.8
private-LUAD	DSMIL	30.8	0.2	17.2	44.8
	MS-DA-MIL	37.4	1.6	10.6	43.4
	MSAA-Net(ours)	40.4	3.4	7.6	41.6

Table 5. Summary of results from conventional multiscale and proposed methods on two datasets. This table is from the paper [20].

Dataset	Method	F1	Precision	Recall
TCGA-LUAD	DSMIL	$\mathbf{0.973 \pm 0.004}$	0.973 ± 0.011	$\mathbf{0.973 \pm 0.012}$
	MS-DA-MIL	0.968 ± 0.008	0.970 ± 0.011	0.966 ± 0.014
	MSAA-Net(ours)	0.970 ± 0.007	$\mathbf{0.975 \pm 0.016}$	0.966 ± 0.014
private-LUAD	DSMIL	0.774 ± 0.084	$\mathbf{0.994 \pm 0.011}$	0.642 ± 0.114
	MS-DA-MIL	0.857 ± 0.043	0.963 ± 0.034	0.779 ± 0.082
	MSAA-Net(ours)	$\mathbf{0.881 \pm 0.031}$	0.928 ± 0.062	$\mathbf{0.842 \pm 0.039}$

of false negatives. Classifying the WSIs from the private-LUAD dataset presents more challenges than those from the TCGA-LUAD dataset due to the smaller cancerous areas in the private-LUAD WSIs, as shown in Fig. 5. Our proposed approach, designed to handle multi-scale WSIs, showed impressive accuracy rates even when faced with challenging data.

Table 4 shows the mean values for TP, FP, FN, and TN. Meanwhile, Table 5 shows the means and standard deviations of F1 score, precision, and recall obtained by five-fold cross-validation for both the proposed and conventional multi-scale methods. In the result of the TCGA-LUAD dataset, the mean number of misclassified WSIs (represented by the sum of FP and FN in Table 4) was 4.8 for DSMIL, 5.6 for MS-DA-MIL, and 5.2 for MSAA-Net. While slight differences resulted, all methods effectively classified the TCGA-LUAD dataset.

In the analysis of the private LUAD dataset, the F1 score achieved by MSAA-Net surpassed that of the conventional methods. Specifically, the proposed method outperformed the DSMIL by 10.7%. In addition, the MSAA-Net exhibited a 20% increase in recall over the DSMIL and a 6.3% improvement over the MS-DA-MIL. According to these results, the two-stage feature aggregation mechanism for multi-scale WSI used in our network contributes to the reduction of oversights compared to the conventional methods.

5 Discussion

This section discusses the performance of attention mechanisms in the MSAA-Net by comparing with those of the conventional methods, namely DSMIL [12] and MS-DA-MIL network [8]. To make the analysis easier to understand, we trained the models using the WSIs containing more than 15% cancerous region for cancer images in the training dataset. Therefore, the number of training examples was 548 (cancer: 189, benign: 359).

5.1 Visualization of Attention Maps

Figures 6 and 7 show the WSIs from the test subset of the private-LUAD dataset alongside the attention maps derived from each method. The top row of images shows a WSI with cancerous regions delineated by a dashed green outline. The subsequent images represent the attention maps generated by the DSMIL, MS-DA-MIL, and MSAA-Net. The prediction results and their probability are shown above the attention maps. The attention maps indicate important regions for classification according to brightness, with brighter areas having a stronger impact on classification. The maps were created by overlaying the brightness corresponding to the attention weights calculated by each method on the image patches. Note that due to the different feature aggregation techniques used by each network, DSMIL presents an attention map for each region, while both the MS-DA-MIL network and the MSAA-Net present attention maps for each region at different scales. Moreover, the attention weight for each scale calculated by MSAA-Net was shown at the bottom of the figure.

Figure 6 shows WSIs containing cancer regions. The WSI shown in Fig. 6 (a) was correctly predicted as a cancer class with high probability by all methods, while the WSI shown in (b) was misclassified by DSMIL. MSAA-Net assigned a high value to the cancerous regions compared to those of the conventional methods for samples (a) and (b). Especially, MSAA-Net correctly located the cancerous region in sample (b). In contrast, attention weight produced by the DSMIL and MS-DA-MIL network could not locate the cancerous region, which resulted in misclassification or low probability prediction, even if correct.

Figure 7 shows WSIs of benign tissues. The WSI shown in Fig. 7 (c) was correctly predicted as a benign class with high probability by all methods, while the WSI shown in (d) was misclassified by the MS-DA-MIL network. For these samples, the proposed network assigned attention weights uniformly to the entire tissue. MSAA-Net independently confirmed each scale whether there were no cancerous regions or not. In contrast, DSMIL uniformly assigned attention weights to the sample (c), similar to the MSAA-Net. However, it focused on a part of the area in sample (d) and predicted it as benign, which reduced its probability. MS-DA-MIL network can focus on multiple images simultaneously. However, it misclassified the sample (d) as cancer. We believe that the misclassification was caused by focusing on a specific region, which reduced the relative importance of the other region.

5.2 Distribution of Attention Weights of Each Method

Figure 8 shows the mean standard deviation of attention weights assigned to tissue images at each scale. Since the standard deviation represents the variability of the attention values, the standard deviation is larger when the attention is assigned locally. In contrast, if the attention is assigned uniformly, the standard deviation will be small because the values have little variation.

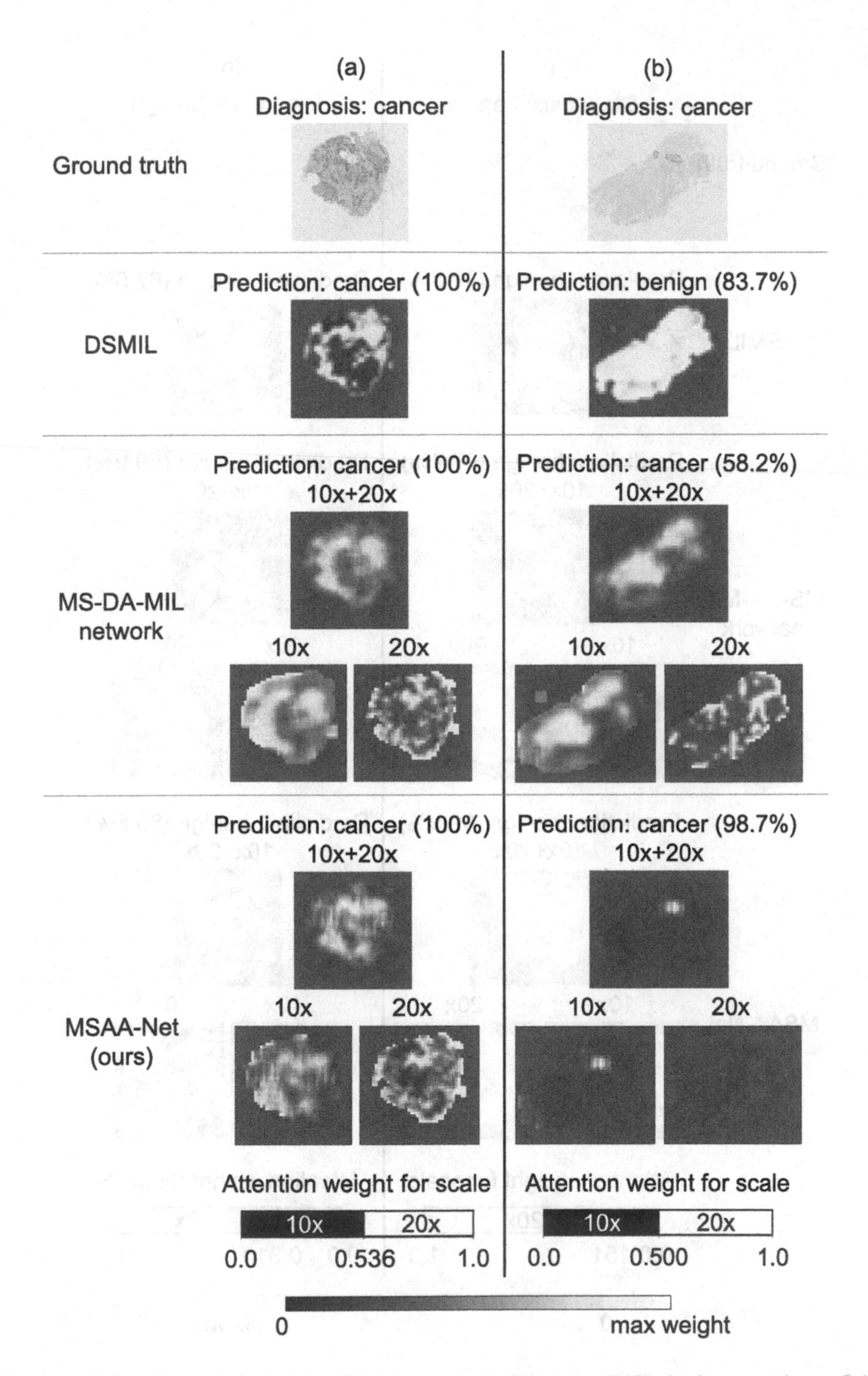

Fig. 6. Attention maps derived from different methods for cancer WSIs in the test subset of the private-LUAD dataset.

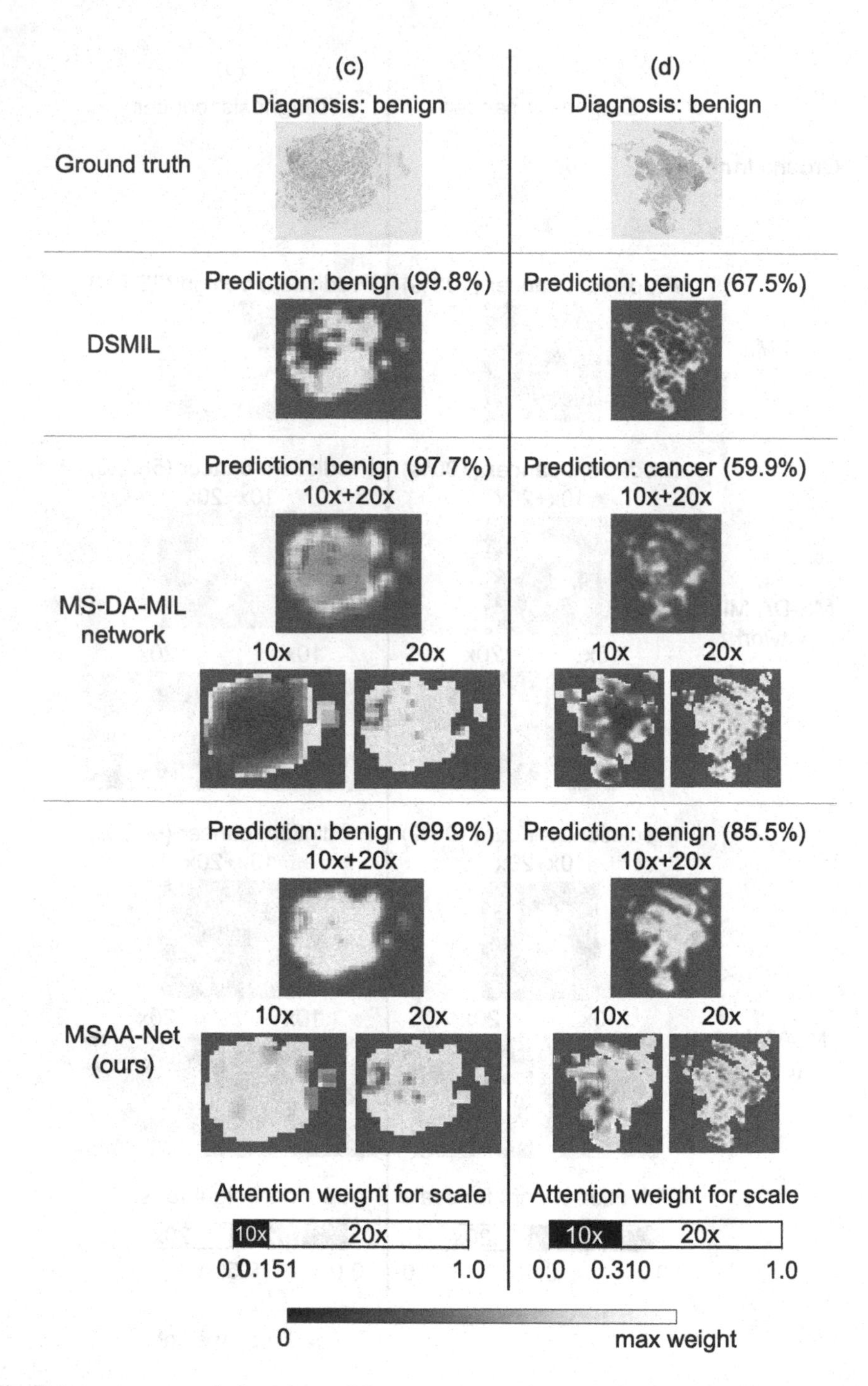

Fig. 7. Attention maps derived from different methods for benign WSIs in the test subset of the private-LUAD dataset.

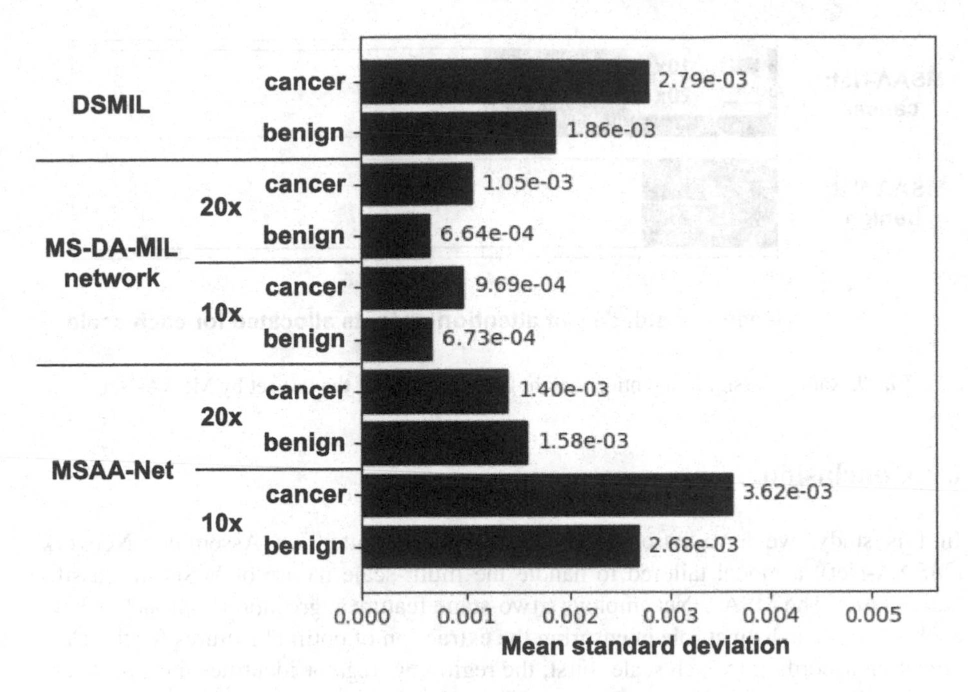

Fig. 8. The mean standard deviation of the attention weight produced by each method.

The standard deviation of the attention maps obtained from MSAA-Net was small for images at 20x magnification and large for images at 10x magnification. It is particularly large for the cancer class at 10x magnification. This indicates that MSAA-Net tends to observe a large area at 20x magnification and focuses on a local region at 10x magnification. For example, in the sample (b) in Fig. 6, MSAA-Net focused on a small cancer region at 10x magnification.

The standard deviation of attention weights produced by DSMIL and MS-DA-MIL network for cancerous WSIs is higher than that for benign WSIs. This means that both methods focus on specific regions for cancerous WSIs. However, the standard deviation of the attention weight for each scale produced by the MS-DA-MIL network is almost the same. We believe that the network failed to capture the characteristics between the different scales of the WSIs.

5.3 Attention Weight Assignment for Each Scale of MSAA-Net

Figure 9 shows the mean and standard deviation of the attention weight for each scale assigned by MSAA-Net. For the classification of the cancer WSIs, the average attention weight assigned for images at 10x magnification was 42%, while the average weight assigned for images at 10x magnification for the classification of benign WSIs was 19%. This suggests that MSAA-Net observes both scales equally for cancer tissues. In contrast, MSAA-Net especially focused on images at 20x magnification for benign tissues. We believe that MSAA-Net achieved accurate classification by adaptively changing the observation magnification.

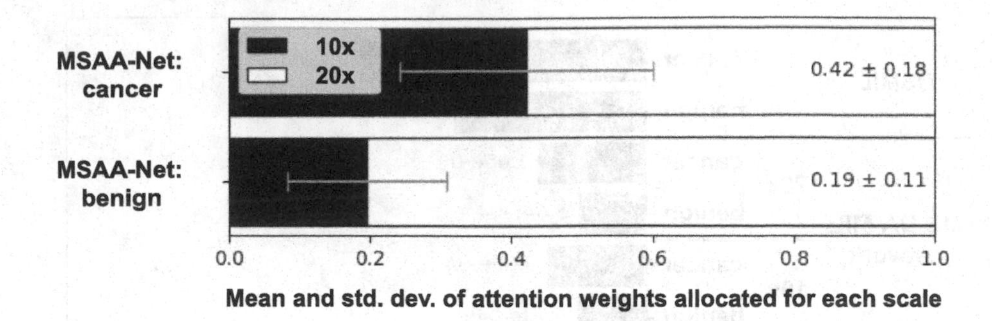

Fig. 9. Ratio of assigned attention weight for each scale of the test set by MSAA-Net.

6 Conclusion

In this study, we have introduced the Multi-Scale Attention Assembler Network (MSAA-Net), a model tailored to handle the multi-scale nature of WSIs in classification tasks. The MSAA-Net employs a two-stage feature aggregation approach, where each stage has a distinct role in ensuring the extraction of optimal features for the classification according to each scale. First, the region aggregator identifies the significant areas in the image at each magnification. Then, the scale aggregator refines the weight for each magnification to improve the classification accuracy.

We verified the proposed method on two different datasets of pathological images of lung tissue. One is based on surgical material published in the TCGA, and the other is based on private biopsy material provided by a cooperating university hospital. The result showed that the MSAA-Net outperformed the state-of-the-art multi-scale methods, particularly in the challenging task of classifying WSIs derived from biopsy materials. Furthermore, by analyzing the visualization of the attention maps, we verified that the feature aggregation mechanism of the MSAA-Net adequately handles multi-scale WSIs.

The proposed multi-scale approach, which utilizes MIL, can automatically adjust the observation magnification according to each case without requiring detailed annotation. We believe that this will enhance the efficiency of automated pathological image analysis.

Acknowledgements. The authors thank Prof. Junya Fukuoka and Dr. Wataru Uegami from Nagasaki University Graduate School of Biomedical Sciences for providing the dataset and medical comments. Computational resource of AI Bridging Cloud Infrastructure (ABCI) provided by the National Institute of Advanced Industrial Science and Technology (AIST) was used. This study is based on results obtained from the project JPNP20006, commissioned by the New Energy and Industrial Technology Development Organization (NEDO). This study was approved by the Ethics Committee (Institutional Review Board) of Nagasaki University Hospital (No. 19081929-2) and the National Institute of Advanced Industrial Science and Technology (No. Hi2019-312) and complied with all the relevant ethical regulations. The results here are in whole or part based upon data generated by the TCGA Research Network: https://www.cancer.gov/tcga.

References

1. Albertina, B., et al.: The cancer genome atlas lung adenocarcinoma collection [TCGA-LUAD]. The Cancer Imaging Archive (2016). https://doi.org/10.7937/K9/TCIA.2016.JGNIHEP5
2. Andrews, S., Tsochantaridis, I., Hofmann, T.: Support vector machines for multiple-instance learning. Adv. Neural Inf. Process. Syst. **15**, 577–584. MIT Press (2002)
3. Araújo, T., et al.: Classification of breast cancer histology images using convolutional neural networks. PLoS ONE **12**(6), e0177544 (2017)
4. Campanella, G., et al.: Clinical-grade computational pathology using weakly supervised deep learning on whole slide images. Nat. Med. **25**(8), 1301–1309 (2019)
5. Chen, R.J., et al.: Scaling vision transformers to gigapixel images via hierarchical self-supervised learning. In: Proceedings of the IEEE Conference on Computer Vision and Pattern Recognition, pp. 16144–16155. IEEE (2022)
6. Dietterich, T.G., Lathrop, R.H., Lozano-Pérez, T.: Solving the multiple instance problem with axis-parallel rectangles. Artif. Intell. **89**(1), 31–71 (1997)
7. Feng, J., Zhou, Z.H.: Deep MIML network. In: Proceedings of the Thirty-First AAAI Conference on Artificial Intelligence, pp. 1884–1890. MIT Press (2017)
8. Hashimoto, N., et al.: Multi-scale domain-adversarial multiple-instance CNN for cancer subtype classification with unannotated histopathological images. In: Proceedings of the IEEE Conference on Computer Vision and Pattern Recognition, pp. 3852–3861. IEEE (2020)
9. Herrera, F., et al.: Multi-instance Regression. In: Multiple Instance Learning, pp. 127–140. Springer, Cham (2016). https://doi.org/10.1007/978-3-319-47759-6_6
10. Ilse, M., Tomczak, J., Welling, M.: Attention-based deep multiple instance learning. In: Proceedings of the 35th International Conference on Machine Learning, vol. 80, pp. 2127–2136. PMLR (2018)
11. Kingma, D.P., Ba, J.: Adam: a method for stochastic optimization. In: 3rd International Conference on Learning Representations, ICLR 2015, San Diego, CA, USA, 7–9 May 2015, Conference Track Proceedings (2015)
12. Li, B., Li, Y., Eliceiri, K.W.: Dual-stream multiple instance learning network for whole slide image classification with self-supervised contrastive learning. In: Proceedings of the IEEE Conference on Computer Vision and Pattern Recognition, pp. 14318–14328. IEEE (2021)
13. Li, J., et al.: A multi-resolution model for histopathology image classification and localization with multiple instance learning. Comput. Biol. Med. **131**, 104253 (2021)
14. Liu, Y., et al.: Detecting cancer metastases on gigapixel pathology images. arXiv:1703.02442 (2017)
15. Maron, O., Lozano-Pérez, T.: A framework for multiple-instance learning. Adv. Neural Inf. Process. Syst. **10**, 570–576. MIT Press (1997)
16. Pinheiro, P.O., Collobert, R.: From image-level to pixel-level labeling with convolutional networks. In: Proceedings of the IEEE Conference on Computer Vision and Pattern Recognition, pp. 1713–1721. IEEE (2015)
17. Shao, Z., et al.: TransMIL: transformer based correlated multiple instance learning for whole slide image classification. Adv. Neural Inf. Process. Syst. **34**, 2136–2147. MIT Press (2021)
18. Simonyan, K., Zisserman, A.: Very deep convolutional networks for large-scale image recognition. In: Proceedings of the International Conference on Learning Representations (2015)
19. Tellez, D., Litjens, G., van der Laak, J., Ciompi, F.: Neural image compression for gigapixel histopathology image analysis. IEEE Trans. Pattern Anal. Mach. Intell. **43**(2), 567–578 (2021)

20. Yoshida, T., Uehara, K., Sakanashi, H., Nosato, H., Murakawa, M.: Multi-scale feature aggregation based multiple instance learning for pathological image classification. In: Proceedings of the 12th International Conference on Pattern Recognition Applications and Methods - ICPRAM, pp. 619–628. Scitepress (2023)
21. Zhouhan, L., et al.: A structured self-attentive sentence embedding. In: 5th International Conference on Learning Representations, ICLR 2017, pp. 24–26 (2017)

Analysis of Generative Data Augmentation for Face Antispoofing

Jarred Orfao and Dustin van der Haar(✉)

Academy of Computer Science and Software Engineering, University of Johannesburg, Kingsway Avenue and University Rd, Auckland Park, Johannesburg, South Africa
{jarredo,dvanderhaar}@uj.ac.za

Abstract. As technology advances, criminals continually find innovative ways to gain unauthorised access, increasing face spoofing challenges for face recognition systems. This demands the development of robust presentation attack detection methods. While traditional face antispoofing techniques relied on human-engineered features, they often lacked optimal representation capacity, creating a void that deep learning has begun to address in recent times. Nonetheless, these deep learning strategies still demand enhancement, particularly in uncontrolled environments. In this study, we employ generative models for data augmentation to boost the face antispoofing efficacy of a vision transformer. We also introduce an unsupervised keyframe selection process to yield superior candidate samples. Comprehensive benchmarks against recent models reveal that our augmentation methods significantly bolster the baseline performance on the CASIA-FASD dataset and deliver state-of-the-art results on the Spoof in the Wild database for protocols 2 and 3.

Keywords: Face antispoofing · Generative data augmentation · Keyframe selection · Analysis

1 Introduction

Facial recognition, as a physical biometric modality, has historically faced challenges regarding user acceptance due to its non-contact nature [19]. However, recent technological advancements and the implementation of COVID-19 protocols have contributed to the increased adoption of facial recognition as a user authentication method in various public and private settings [5]. This includes its integration into workplaces, train stations, and airports, where computers and mobile phones are the primary authentication devices. Despite significant progress in facial recognition, it remains vulnerable to inherent weaknesses in its components. Among these components, the biometric sensor is the first point of interaction for users, making it relatively accessible and less expensive to exploit. Unlike other system elements, the sensor lacks control over the input it receives, rendering it susceptible to presentation attacks such as face spoofing.

Face spoofing refers to an attack where an unauthorised individual, known as an attacker, attempts to deceive the biometric sensor by presenting a two-dimensional medium (e.g., a photo or video) or a three-dimensional medium (e.g., a mask) of a

M. De Marsico et al. (Eds.): ICPRAM 2023, LNCS 14547, pp. 69–94, 2024.
https://doi.org/10.1007/978-3-031-54726-3_5

registered user, commonly referred to as the *victim* [13]. Since an authenticated face is often the primary barrier to accessing physical and digital assets, detecting and preventing face spoofing incidents is crucial.

In this research, we concentrate on elucidating the relationship between variability and the fidelity of images generated, specifically focusing on its implications for face antispoofing performance. Building upon our prior work [33], which explored the potential of Generative Adversarial Networks (GANs) for data augmentation in face antispoofing, this extension delves deeper by offering quantitative and qualitative analyses of the generated image fidelity. Moreover, we introduce a t-distributed Stochastic Neighbour Embedding (t-SNE) plot to clarify the models' predictions for each Spoof in the Wild (SiW) protocol. The main findings of this research are summarised accordingly:

1. We show the effectiveness of GANs as a data augmentation strategy for face antispoofing compared to traditional augmentation approaches.
2. We propose an unsupervised keyframe selection process for more effective candidate generation and investigate the relationship between variability and image fidelity and its role in artefact detection.
3. We explore when data augmentation should be performed, the optimal data augmentation percentage and the number of frames to consider for face antispoofing.
4. We conduct a comprehensive benchmarking analysis, comparing our proposed approach against the current state-of-the-art face antispoofing methods using public datasets.

The remaining sections of the paper are organised as follows: Sect. 2 provides a comprehensive discussion on face antispoofing and related research. Section 3 presents a detailed explanation of the proposed method. Section 4 outlines the experimental setup employed in this study. Section 5 provides the experiment results and their analysis, followed by the concluding remarks in Sect. 6.

2 Similar Work

For more than 15 years, face antispoofing has been a prominent area of research, with an increasing number of publications every year [51]. However, even with extensive efforts from researchers, face recognition systems still face vulnerabilities to straightforward and non-intrusive attack vectors. These attack vectors exploit weaknesses in biometric sensors and can be classified accordingly [16]:

1. A *photo attack* involves an attacker presenting a printed image of a *victim* to the biometric sensor.
2. A *warped-photo attack* is an extension of a photo attack where the printed image is manipulated to simulate facial motion.
3. A *cut-photo attack* is an extension of a photo attack where the printed image has eye holes removed, creating the illusion of blinking.
4. A *video replay attack* occurs when an attacker replays a video containing a *victim's* face to the biometric sensor.

5. A *3D-mask attack* is when an attacker wears a 3D mask that replicates a *victim's* facial features in front of the biometric sensor.
6. A *DeepFake attack* involves an attacker using deep learning methods to replace a person's face in a video with a victim's face.

In recent years, significant advancements have been made in face antispoofing methods. Traditionally, researchers relied on human vitality cues and handcrafted features to address this challenge. Vitality cues such as eye blink detection [25], face movement analysis [3], and gaze tracking [1] were commonly used. However, these approaches proved to be vulnerable to cut-photo and video-replay attacks, rendering them unreliable. On the other hand, handcrafted features like Local Binary Patterns (LBP) [8], and Shearlets [15] have shown effectiveness in extracting spoof patterns in real-time with minimal resource requirements. These features have demonstrated their capability to detect spoofing attempts. However, the drawback of handcrafted features lies in the need for feature engineers to manually select essential features from images, which becomes more challenging as the number of classification classes increases [41]. Additionally, each feature requires meticulous parameter tuning, adding to the complexity of the process.

In contrast, deep learning approaches offer the advantage of independently discovering descriptive patterns with minimal human intervention. Convolutional Neural Network (CNN) architectures have been successfully applied in face antispoofing. However, they often require a substantial amount of training data to ensure effective model performance and are susceptible to overfitting. To mitigate overfitting during training, researchers employ regularization techniques such as dropout, particularly when training a model without prior knowledge [40].

Another strategy that has shown promise is leveraging pre-trained models and fine-tuning specific layers for the task at hand [14,32]. This approach enables the model to utilize the learned features from a large dataset and adapt them to a similar task with a smaller dataset. Some researchers have achieved notable results by incorporating auxiliary information into CNN training. For instance, in the work of [30], competitive performance was attained by fusing the depth map of the last frame with the corresponding remote photoplethysmography signal obtained across a sequence of frames to determine the final spoof score. However, this approach has limitations due to its reliance on multiple frames, which may restrict its applicability in certain scenarios.

In recent advancements, researchers have made significant progress in face anti-spoofing by adopting a hybrid approach that combines handcrafted features with deep learning techniques. For instance, in the study conducted by Wu et al. [45], they introduced a DropBlock layer that randomly discards a portion of the feature map. This approach enables the learning of location-independent features and serves as a form of data augmentation by mimicking occlusions. By incorporating blocked regions, the training samples are increased, reducing the risk of overfitting and enhancing the model's performance. Likewise, drawing inspiration from the descriptive nature of Local Binary Patterns (LBPs) in capturing local relations, Yu et al. [53] proposed a novel Central Difference Convolution layer. This layer retains the same sampling step as a traditional convolutional layer but emphasizes the aggregation of center-oriented gradients from the sampled values. This technique allows for the extraction of both intensity-level semantic information and gradient-level detailed messages, which the researchers demonstrated to be crucial for effective face antispoofing.

The above analysis shows that hybrid approaches reap the benefits of traditional and deep learning methods. Furthermore, approaches such as Wu et al. [45] also act as a data augmentation strategy. To fairly evaluate the effectiveness of our data augmentation strategy, we will fine-tune a vision transformer similar to [14], which we will discuss in the next section.

3 Proposed Method

This paper proposes a transfer learning approach to face antispoofing using a pre-trained Vision Transformer (ViT) model. Furthermore, we use generative data augmentation to optimise this model by synthesising candidate samples using StyleGAN3 models. In the following sections, we will discuss the different stages of the training pipeline, illustrated in Fig. 1.

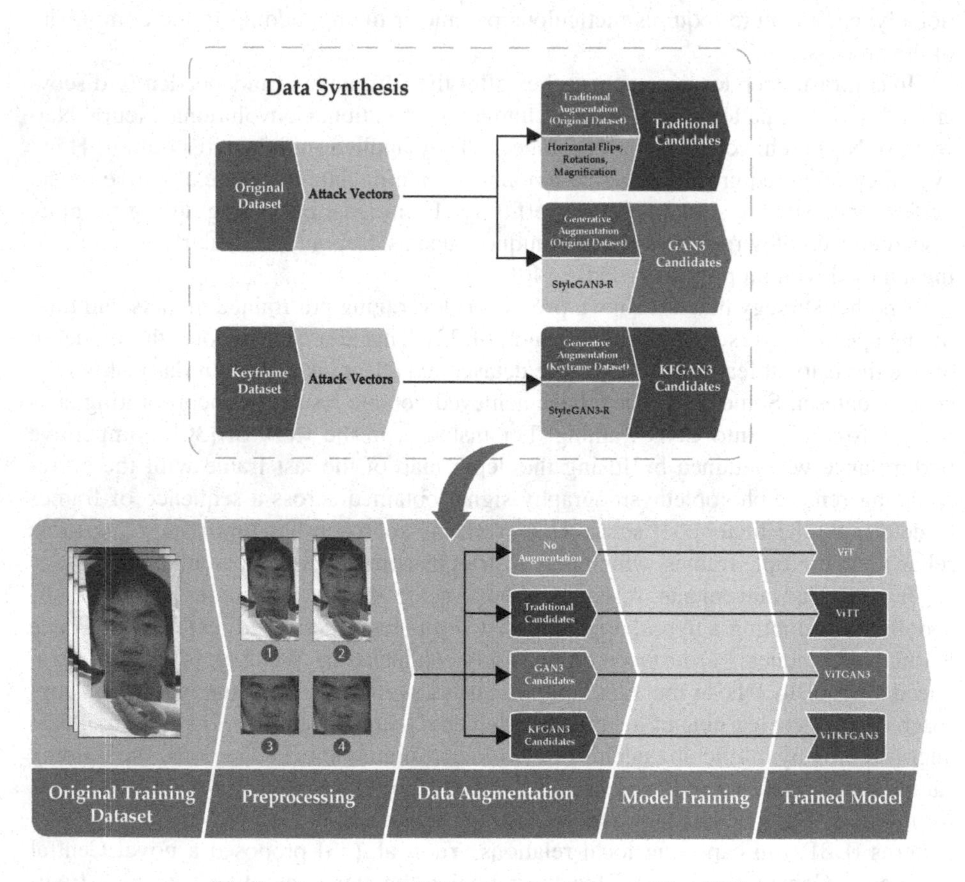

Fig. 1. The stages of the proposed training pipeline [33].

3.1 Preprocessing

To mitigate background and dataset biases, we performed preprocessing on each video. The first step involved utilizing the MTCNN algorithm [54] for accurate face detection. Subsequently, we performed additional transformations on the detected region. This included rotating the region to align the eye centers horizontally, scaling the region to minimize the background, and cropping the region to obtain a square patch that focused on the subject's eyebrows and mouth. While it was possible to scale the detected region to eliminate the background entirely, doing so resulted in the exclusion of eyebrows and the mouth from the crop patch. Considering the significance of these facial features in conveying emotions, we opted to include a slightly larger background area to ensure their presence in the cropped patch.

3.2 Data Augmentation

Instead of applying affine transformations to existing images, we adopted a generative data augmentation approach. For each attack vector, we trained separate StyleGAN3 models, which allowed us to generate new candidate samples. This generative approach provided us with a diverse set of synthesized training images. In addition, we also employed traditional data augmentation methods for comparison with our generative approach. By incorporating both approaches, we aimed to evaluate their respective effectiveness in enhancing the training data.

$$N_I = \left(\frac{N_S \times P}{100\% - P} \right) \div N_A \tag{1}$$

In Eq. (1), N_I is the number of images generated for each attack vector; N_S is the number of samples present in the training protocol; N_A is the number of attack vectors present in the training protocol, and P is the desired data augmentation percentage. Using Eq. 1, we calculated the number of images necessary to achieve the desired data augmentation percentage for each attack vector.

3.3 Model Training

The advent of vision transformers has attracted significant attention, initially being utilised for natural language processing tasks. However, their success in this domain has piqued the interest of computer vision researchers. In a groundbreaking study [23], researchers adapted transformers for computer vision by introducing modifications. Instead of feeding tokens as input, they divided images into patches and employed patch embeddings, giving rise to Vision Transformers (ViTs). ViTs have achieved remarkable performance in image classification tasks [24].

Inspired by the work of George et al. [14], who achieved state-of-the-art results in face antispoofing using ViTs, we adopted a similar approach. In our study, we also consider face antispoofing as a binary classification problem, distinguishing between *bona fide* samples (genuine samples obtained directly from individuals) and *spoof* samples (fabricated samples acquired indirectly from presented mediums). Lastly, we employ a static face antispoofing approach by analysing each frame independently.

4 Experiment Setup

4.1 Datasets

We selected the Casia Face Antispoofing Database (CASIA-FASD) and the Spoof in the Wild (SiW) Database to evaluate our face antispoofing approach.

CASIA-FASD [56] was created in 2012 and contains 600 videos captured in natural scenes with no artificial environment adjustments. It involves fifty subjects who were recorded under normal conditions (N) and subjected to cut-photo (C), warped-photo (W), and video-replay (R) attacks. The attacks were captured using cameras of varying resolutions (low, medium, and high). The dataset provides seven protocols for training and evaluating models involving different resolution and attack types combinations.

SiW [30] was created in 2018 and consists of 4,478 videos involving 165 subjects with variations in distance, pose, illumination, and expression. It includes normal (N) videos, as well as photo (P) and video-replay (R) attack videos. Normal videos were captured with and without lighting variations, and the photo attacks were created using low-resolution (LR) and high-resolution (HR) printed images. Video replay attacks utilise various devices for displaying bona fide videos.

SiW offers three protocols for evaluating models. Protocol 1 assesses generalisation by training on the first 60 frames of the training set and testing on all frames of the test set. Protocol 2 focuses on generalisation across different spoof mediums using a leave-one-out strategy, while Protocol 3 evaluates model performance on unknown presentation attacks using different attack vectors. The protocols consider various combinations of training and testing data. For clarity in later sections, we denoted LOO_X as the group left out for training and used exclusively for testing, where X is a video-replay spoof medium (ASUS, IP7P, IPP2017 or SGS8) in protocol two and an attack vector (P or R) in protocol three.

Since protocols 2 and 3 utilise various training combinations, it is essential to use the mean and standard deviation of the combinations when reporting metrics. For comparability with other work, we used the videos of subjects 90 for training and 75 for testing.

While CASIA-FASD may be considered relatively older and smaller than other face antispoofing datasets, we chose it precisely because it includes cut-photo attacks and low-resolution videos, which are essential to our research. Additionally, we selected SiW due to its extensive variability in subjects' distance, pose, illumination, and expression. These factors provide a more comprehensive and diverse representation of real-world scenarios in our evaluation.

4.2 Dataset Augmentation

The traditional method of data augmentation involves applying random transformations to an image, such as adjusting brightness, rotation, or cropping [35]. However, in certain problem environments, these transformations can have an adverse impact on the sample's label [38]. For example, in the case of video-replay attack samples, the backlight of an LCD screen can make them appear brighter compared to genuine samples, as illustrated in Fig. 9. Consequently, altering the brightness of spoof samples may introduce label inconsistencies and result in an unclear decision boundary. Therefore, when employing the traditional data augmentation approach, we only utilize random horizontal flips, rotations (within 15°), and magnifications (within 20%).

In contrast to traditional data augmentation techniques, Generative Adversarial Networks (GANs) offer the ability to generate new samples that exhibit the characteristics of a specific image domain while preserving the labels assigned to the original samples. For our generative data augmentation approach, we specifically chose the StyleGAN3 architecture. StyleGAN3 represents a significant advancement in image synthesis by associating image details with object surfaces rather than absolute coordinates [21], which enables it to achieve impressive results even with smaller datasets. This capability allows StyleGAN3 to generate high-fidelity images with limited data, positioning it as the state-of-the-art model in image generation.

We utilised the alias-free rotation equivariant architecture known as StyleGAN3-R to generate candidate samples. Each model was trained using the recommended configuration, utilising 2 GPUs and a resolution of 256 by 256 pixels, for a total of 5000 kimg. In this context, *kimg* refers to the number of thousand images sampled from the training set. To select the most suitable model, we evaluated the models based on their Frechet Inception Distance (FID) [17], ultimately choosing the model with the lowest FID. For the generation process, we employed ordered seeds ranging from 1 to N_I, and a truncation psi value of 1 to maximise variation in the generated samples.

For clarity, we synthesised images for each attack vector separately, with each image labelled as a spoof. Although it is possible to generate images using bona fide samples, labelling them as such could introduce a DeepFake attack vulnerability. Figure 2 illustrates the synthesised images for each SiW attack vector using each data augmentation approach.

4.3 Face Antispoofing

We employed transfer learning to alleviate the computational burden of training a Vision Transformer (ViT) from the ground up. This approach harnesses the knowledge embedded within a pre-trained model from one domain and repurposes it for a different task. Guided by the procedure delineated by George et al. [14], we utilised a ViT-B/32 model initially trained on ImageNet, adapting it further to meet our needs. We resized our input images to a 224×224 pixel frame. The model's final layer was swapped out for a dense layer containing two nodes, which we identify as the classification layer, and subsequently activated using a SoftMax function. All layers were set to non-trainable except for the classification layer. Our training regimen incorporated binary cross-entropy loss, optimised through the Adam optimizer [22], using a learning rate set at 1e−4. We implemented early stopping with a patience value of 15 to prevent overfitting and trained each model for 70 epochs. We chose the model iteration with the minimal validation loss to partake in the testing procedure.

To identify the best data augmentation percentage, we conducted a hyperparameter search [29], testing augmentation percentages of 5, 10, 20, and 30. Leveraging a stratified 3-fold cross-validation strategy [36], the training dataset was divided, with 80% allocated for training and the remaining 20% for validation; each segment

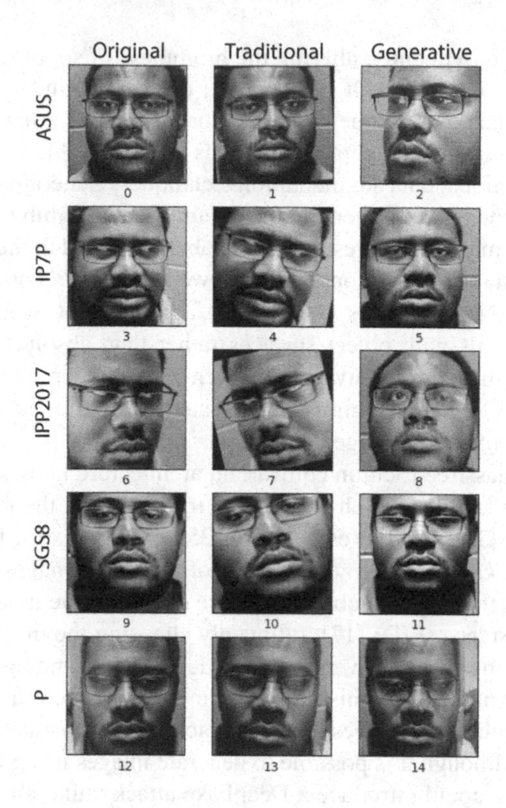

Fig. 2. The synthesised images for each SiW attack vector. The columns and rows correspond to the sample sets and attack vectors [33].

was handled in batches of 32. This process was repeated twice. For performance assessment and comparison with related works, we used the average values of several $\text{ISO/IEC}_{30107-3}$ [18] metrics:

1. Attack Presentation Classification Error Rate (APCER): The proportion of attack samples mistakenly recognised as genuine.
2. Bonafide Presentation Classification Error Rate (BPCER): The proportion of genuine samples mistakenly labelled as an attack.
3. Average Classification Error Rate (ACER): The average of APCER and BPCER.

We also included the Equal Error Rate (ERR) to facilitate comparisons with earlier face antispoofing methods. Within biometric antispoofing, ERR represents the juncture at which APCER and BPCER coincide [4].

4.4 Keyframe Selection

Minimal movement in a one-second video can produce up to 24 almost identical frames. We delved into understanding how these near-replica frames influence the ViT models' training. Our approach comprised a three-part unsupervised keyframe selection procedure:

Fig. 3. The original frames for subject 18, video 1 in the CASIA-FASD test release [33].

Stage 1: Feature Extraction. Using a ResNet-50 backbone that's been pre-trained on the VGGFace2 dataset, we extracted features from the pre-processed frames. The VGGFace2 dataset, rich with 9131 diverse subjects showcasing varied ages, ethnicities, and professions under different lightings and angles [10], was appropriate for our objective.

Stage 2: Feature Clustering. We grouped the features we extracted through Lloyd's K-Means clustering technique, facilitated by the Facebook AI Similarity Search (FAISS) [20] library. To pinpoint the optimal value for K, we conducted a hyperparameter search to optimize the Silhouette Score [37].

Stage 3: Keyframe Selection. For each category, we calculated the mean optimal K. For CASIA-FASD, we used attack vectors as our categorization criteria. In contrast, for Spoof in the Wild, we based our categories on the media labels associated with each session. We then re-clustered the features using the mean optimal K specific to each category to derive cluster centroids. Vector quantization helped us pinpoint the features nearest to these centres, allowing us to select the related images as the keyframes.

Figures 3 and 4 respectively illustrate the keyframe selection process before and after its application. Table 1 shows the ablation study for selecting the optimal K for CASIA-FASD. It's evident from the table that the standard deviation exceeds the mean, indicating a notable spread in optimal Ks, particularly between the 75th and 100th percentile values. Additionally, the 75th percentile values hover near the mean except for one category: the medium resolution warped-photo attack (W_2). This suggests that a

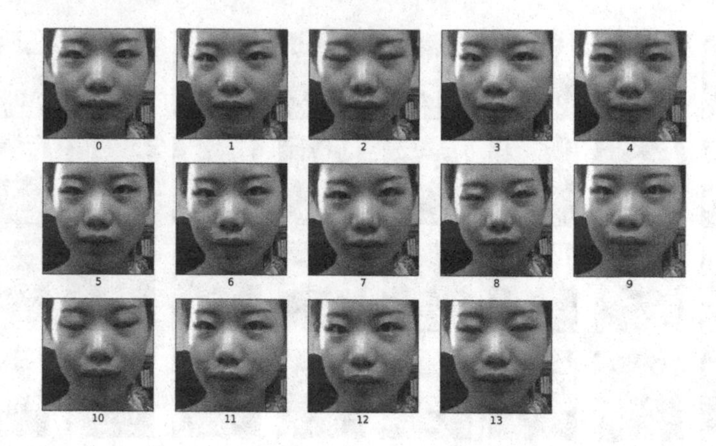

Fig. 4. The keyframes for subject 18, video 1 in the CASIA-FASD test release [33].

category-specific mean optimal K is preferable over an individual video's due to certain outlier videos. The impact of keyframe reduction on CASIA-FASD is depicted in Fig. 5. This unsupervised reduction led to a 1412% frame decrease, narrowing down 110882 original frames to just 7850 keyframes.

To simplify the narrative, we've implemented particular notations that differentiate between the Vision Transformer models depending on their data augmentation methods. The foundational model, devoid of data augmentation and trained on the original dataset, is denoted as ViT. The model enhanced using conventional data augmentation is labelled ViTT. Given our generative data augmentation uses StyleGAN3-R models, we designate a ViT model adapted with this method as ViTGAN3. We introduce KFGAN3, a keyframe-centric generative data augmentation method that employs StyleGAN3 models, which are exclusively trained on keyframes. Hence, a ViT model leveraging this keyframe-centric augmentation is labelled ViTKFGAN3.

Table 1. The mean, std. deviation and five-number-summary for the optimal K of each attack vector in CASIA-FASD [33].

Attack Vector	Five-Number-Summary					
	Mean	Min	25th	50th	75th	Max
N_1	14.2 ± 15	2	2	5	23.3	42
N_2	9.6 ± 16.8	2	2	2.5	4.3	54
N_HR	12.3 ± 20.2	2	2	3	5.3	68
C_1	2.2 ± 0.2	2	2	2	2	3
C_2	2.2 ± 0.5	2	2	2	2	4
C_HR	3.1 ± 4.2	2	2	2	2	21
W_1	32.8 ± 21.4	2	20.3	31.5	49.5	73
W_2	17.4 ± 30.9	2	2	3	4.5	95
W_HR	20.7 ± 28.5	2	2.8	4.5	39.8	101
R_1	14.1 ± 20.8	2	2	2.5	18.3	63
R_2	10.8 ± 13	2	2	2	21.3	40
R_HR	18.2 ± 29.5	2	2	2.5	14.5	85

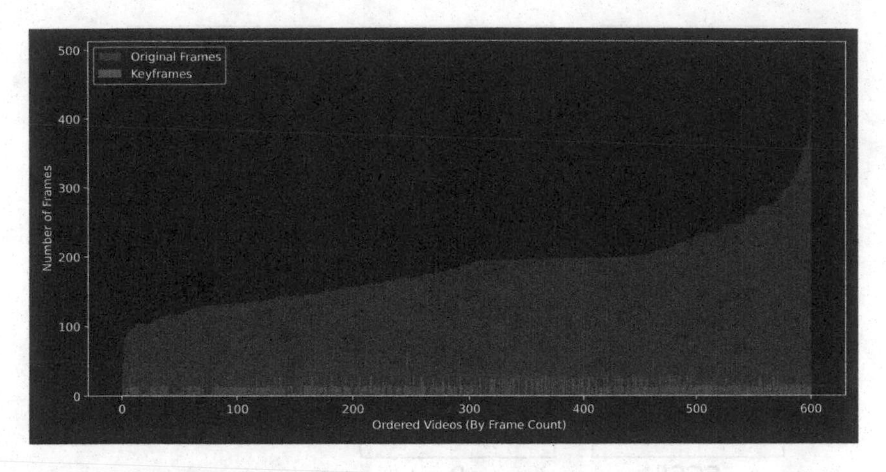

Fig. 5. The number of original frames (blue) vs the number of keyframes (red) for each video in the CASIA-FASD ordered by video frame count [33]. (Color figure online)

5 Results

In this section, we examine the results of the StyleGAN3 models regarding the quality of the generated candidate samples and the effectiveness of generative data augmentation for face antispoofing. Furthermore, we conducted an ablation study using SiW protocols 2 and 3 to determine when to perform data augmentation, the optimal data augmentation percentage, and the optimal number of frames to detect face spoofing.

Generated Candidate Sample Quality. Figure 6 showcases smoother curves compared to Fig. 7, indicating a relatively easier training process when using all the frames compared to only keyframes. This result was expected since keyframe samples exhibit more variations, making it more challenging for the StyleGAN3 models to generate samples that accurately represent the entire distribution of real samples. Additionally, it is worth noting that there are considerably fewer keyframe samples for each attack vector than normal samples, as presented in Table 2. Despite these challenges, the KFGAN3 models converged, albeit with slightly higher FID values than the GAN3 models. This difference can be attributed to the adaptive discriminator augmentation inherited from the StyleGAN2-ADA architecture [21].

Table 2. The number of samples present in each SiW attack vector when training the GAN3 and KFGAN3 models.

Model	ASUS	IP7P	IPP2017	SGS8	P	R
KFGAN3	40	44	48	20	162	151
GAN3	1788	1796	1803	900	1925	6287

Table 3 displays the FID (Fréchet Inception Distance) and kimg values for the selected GAN3 and KFGAN3 models trained on the SiW dataset. The FID metric

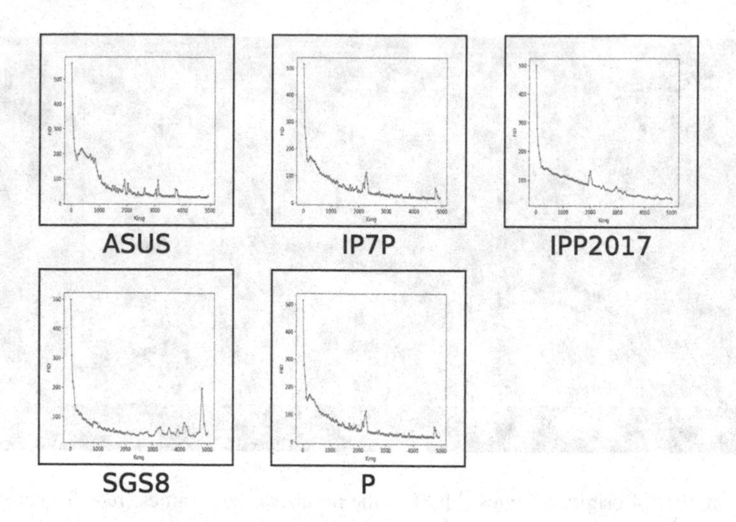

Fig. 6. The training progression of the SiW GAN3 candidate samples in terms of FID.

Table 3. The FID and kimg of the selected GAN3 and KFGAN3 models for SiW.

Attack Vector	GAN3		KFGAN3	
	FID	kimg	FID	kimg
ASUS	25.33	4840	54.66	4160
IP7P	18.18	4680	57.91	4200
IPP2017	34.76	5000	99.41	4920
SGS8	35.14	4560	82.98	4240
P	26.47	4880	66.38	3440

quantifies the similarity between the generated and real samples, with lower values indicating higher quality. According to the FID values, the IP7P GAN3 models produce the highest quality samples, closely resembling the real samples. Conversely, the IPP2017 KFGAN3 models generate samples of the lowest quality. Figure 9 visually compares the generated samples with the real samples, demonstrating a substantial fidelity between them. However, some anomalies may contribute to the slightly higher FID values observed. Nevertheless, the overall performance of the StyleGAN3 models is satisfactory, as they effectively generate high-quality and realistic candidate samples.

Although the FID metric is a useful indicator of sample quality, our investigation revealed that the lowest FID value did not consistently correspond to the best-generated samples. As depicted in Fig. 8, the SGS8 samples generated by the KFGAN3 model with the lowest FID value (81.40) exhibit lower contrast compared to the real samples. Therefore, while we employed the FID as a guiding factor for selecting the StyleGAN3 models, we also conducted manual inspection to make the final decision. Figure 9 illustrates the generated SGS8 samples produced by the manually inspected KFGAN model. Despite having a slightly higher FID value (82.98), the generated samples are more representative of the real samples' image quality.

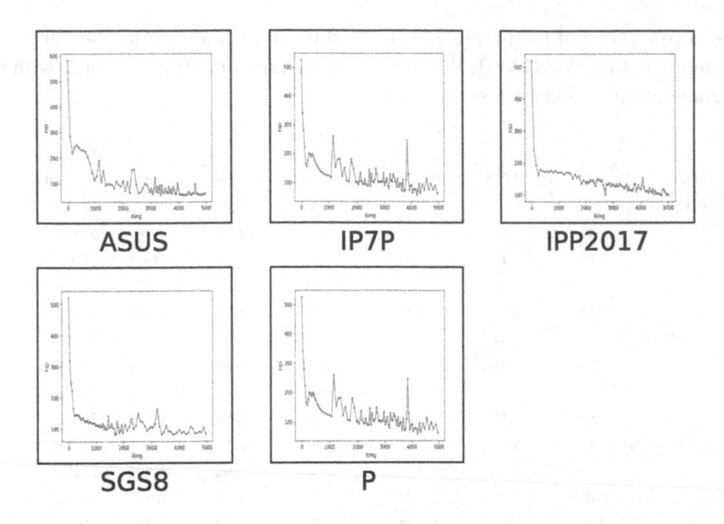

Fig. 7. The training progression of the SiW KFGAN3 candidate samples in terms of FID.

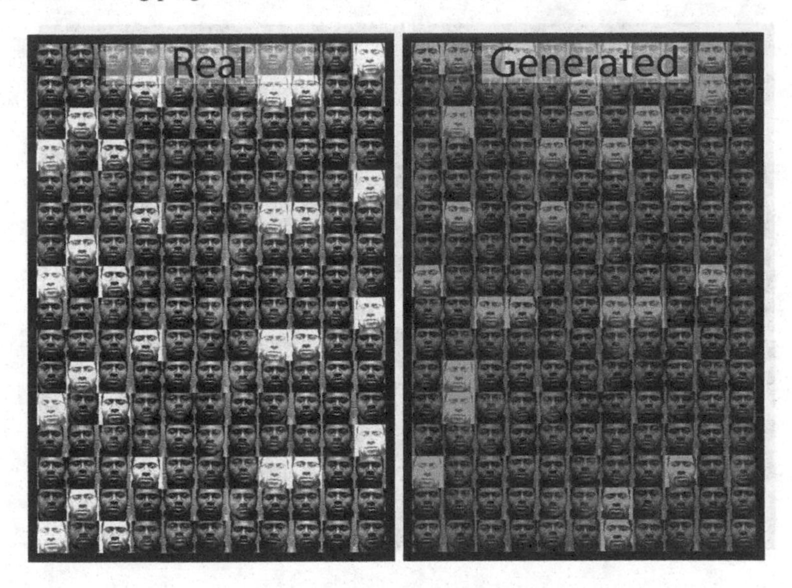

Fig. 8. The SGS8 samples generated by the KFGAN3 model with the lowest FID (81.40).

Data Augmentation Impact on Sample Variability: Our study investigated the sample variability within sets (intra-set) and between sets (inter-set) for SiW protocols 2 and 3 at different data augmentation percentages. The aim was to assess the impact of each data augmentation approach quantitatively. To evaluate the intra-set variability, we computed the average Frechet Inception Distance (FID) between the bona fide and spoof samples within the training set. As for the inter-set variability, we calculated the average FID between the training and testing sets.

Table 4 presents the analysis of intra-set and inter-set sample variability for SiW protocol 2. It indicates that the GAN3 approach leads to increased variability in both intra-

Table 4. The performance of the baseline (ViT), traditional data augmentation (ViTT) and generative data augmentation (ViTGAN3, ViTKFGAN3) models for SiW protocol 2, with the corresponding average intra-set and inter-set variability.

						Variability (FID)	
Model	Aug. (%)	APCER (%)	BPCER (%)	ACER (%)	EER (%)	Intra-set	Inter-set
ViT (Baseline)	0	4,91	2,11	3,51	6,07	36992	217112
ViTT	5	4,96	1,81	3,39	6,01	37467	214457
	10	4,4	2,6	3,5	6,23	38361	212023
	20	3,93	3,39	3,66	6,09	40022	207570
	30	**4,1**	3,5	3,8	6,58	42503	204420
ViTGAN3	5	5,84	2,04	3,94	6,3	36875	218471
	10	4,94	1,64	**3,29**	**5,88**	37529	219449
	20	5,13	2,63	3,88	6,69	38071	221746
	30	5,33	**1,43**	3,38	5,99	38793	223328
ViTKFGAN3	5	5,51	2,42	3,97	6,75	36930	215787
	10	4,99	1,96	3,47	5,98	36793	214515
	20	5,24	2,04	3,64	6,23	36557	212327
	30	5,12	2,35	3,74	6,02	36600	210849

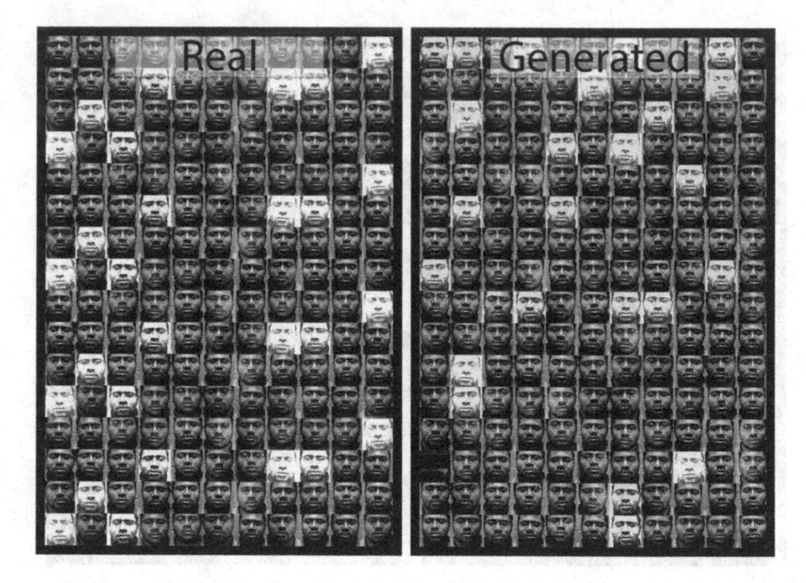

Fig. 9. The SGS8 samples generated by the manually inspected KFGAN3 model, possessing an FID of 82.98.

set and inter-set samples. On the other hand, the Traditional and KFGAN3 approaches result in reduced variability in both cases. Similarly, Table 5 displays the findings for SiW protocol 3. In this case, the Traditional and GAN3 approaches increase the intra-set variability while reducing the inter-set variability. However, the KFGAN3 approach continues to reduce both intra-set and inter-set variability.

Interestingly, the best results were achieved in protocol 2 when both intra-set and inter-set variability increased, while in protocol 3, the optimal outcome was observed when both intra-set and inter-set variability decreased. Based on the nature of these

Table 5. The performance of the baseline (ViT), traditional data augmentation (ViTT) and generative data augmentation (ViTGAN3, ViTKFGAN3) models for SiW protocol 3, with the corresponding intra-set and inter-set variability.

Model	Aug. (%)	APCER (%)	BPCER (%)	ACER (%)	EER (%)	Variability (FID)	
						Intra-set	Inter-set
ViT (Baseline)	0	17,28	**0,55**	8,92	14,03	41 630	207 983
ViTT	5	17,2	0,96	9,08	13,94	46 199	203 888
	10	15,08	0,84	7,96	12,29	46 932	200 823
	20	15,04	0,83	7,94	12,1	49 008	194 687
	30	14,08	1,16	7,62	11,81	50 925	189 946
ViTGAN3	5	17,66	0,6	9,13	13,9	46 579	207 258
	10	23,28	0,68	11,98	17,22	46 797	206 643
	20	21,77	0,74	11,25	16,5	46 715	204 835
	30	21,83	0,96	11,39	17,32	47 010	203 222
ViTKFGAN	5	**13,79**	0,95	**7,37**	**11,5**	46 167	206 377
	10	16,42	2,24	9,33	14,09	46 026	204 858
	20	13,75	1,02	7,38	11,82	45 610	201 331
	30	15,18	1,5	8,34	12,94	45 365	198 544

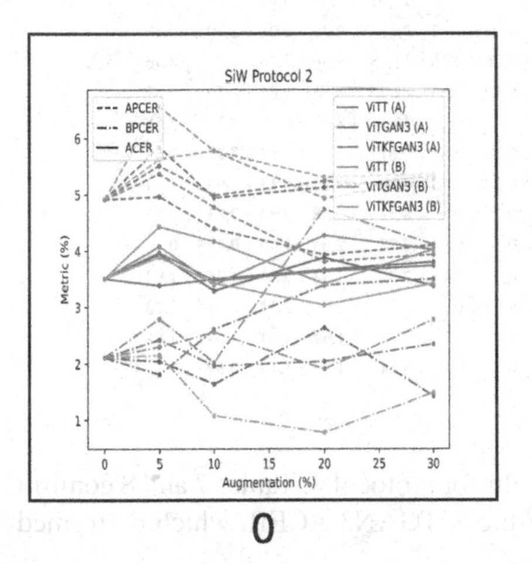

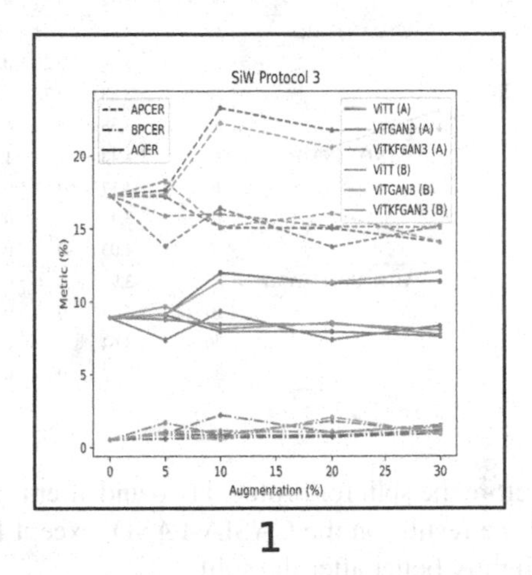

Fig. 10. The average metrics of the traditional (ViTT) and generative (ViTGAN3 and ViTKFGAN3) data augmentation models performed before (B) and after (A) the validation split for SiW protocols 2 and 3, using data augmentation percentages: 5, 10, 20 and 30 [33].

protocols, reducing both intra-set and inter-set variation seems beneficial for unseen presentation attacks (protocol 3) and increasing them seems beneficial for unseen spoof medium attacks of the same type (protocol 2).

Data Augmentation Before or After the Validation Split: As illustrated in Fig. 10, the generative data augmentation approach (GAN3) performed better for both protocols and achieved its lowest ACER before the validation split. In contrast, the traditional data augmentation approach performed better and achieved its lowest ACER after the split. Interestingly, the keyframe generative data augmentation (KFGAN3) performed better

Table 6. Performance of the baseline (ViT), traditional data augmentation (ViTT), and generative data augmentation (ViTGAN3, ViTKFGAN3) models for SiW protocols 2 and 3 in terms of ACER (%) for the image-based and video-based classification approaches, using window sizes of 5, 7, 10, and 15 [33].

		Protocol 2					Protocol 3				
Model	Aug. (%)	Image	5	7	10	15	Image	5	7	10	15
ViT (Baseline)	0	3.51	**0**	**0**	**0**	**0**	8.92	2.83	1.49	1.19	**0**
ViTT (Before)	5	4.08	0.78	0.26	0.26	0	8.79	2.08	2.08	1.19	0.3
	10	3.4	0	0	0	0	8.45	2.9	2.9	2.01	0.3
	20	4.28	0.26	0.26	0.26	0.26	8.48	2.98	2.68	1.49	0
	30	4.02	0.26	0.26	0	0	8.06	2.98	2.98	2.38	0.3
ViTT (After)	5	3.39	0	0	0	0	9.08	3.13	2.53	1.93	1.34
	10	3.5	0	0	0	0	7.96	1.79	1.79	1.49	1.04
	20	3.66	0	0	0	0	7.94	3.27	2.68	1.19	0.3
	30	3.8	0.26	0.26	0.26	0	7.62	2.68	2.68	1.79	0.3
ViTGAN3 (Before)	5	3.89	0	0	0	0	9.12	2.68	2.38	2.08	1.79
	10	3.43	0	0	0	0	11.4	5.21	5.21	4.02	4.32
	20	**3.04**	0.52	0.52	0	0	11.34	3.57	3.57	2.68	2.68
	30	3.42	0	0	0	0	12.01	5.51	5.51	4.61	5.21
ViTGAN3 (After)	5	3.94	1.04	1.04	1.04	0.52	9.13	2.38	2.08	1.79	0.89
	10	3.29	0.26	0.26	0.26	0.26	11.98	5.36	5.36	4.46	4.46
	20	3.88	0.78	0.26	0.52	0.26	11.25	5.95	5.65	4.76	3.87
	30	3.38	0	0	0	0	11.39	8.78	8.78	8.18	7.59
ViTKFGAN3 (Before)	5	4.43	1.3	1.3	0.78	0.52	9.67	3.87	3.87	3.27	2.68
	10	4.17	0.52	0.52	0.52	0.52	8.13	3.13	3.13	1.93	1.04
	20	3.42	0	0	0	0	8.57	2.08	2.83	2.83	2.23
	30	4.03	0	0	0	0	7.73	2.08	1.49	**0.595**	0
ViTKFGAN3 (After)	5	3.97	0.26	0.52	0	0	**7.37**	1.93	2.53	1.34	1.93
	10	3.47	0	0	0	0	9.33	2.9	2.31	2.01	2.23
	20	3.64	0	0	0	0	7.38	**1.64**	1.64	1.64	1.04
	30	3.74	0	0	0	0	8.34	1.79	**1.19**	**0.595**	0

before the split for protocol two and after the split for protocol 3. Tables 7 and 8 confirm these results on the CASIA-FASD, except for the ViTGAN3 ACER, which performed slightly better after the split.

Image-Based Classification Versus Video-Based Classification: One advantage of employing a static analysis approach in face antispoofing is its compatibility with both image-based and video-based recognition systems. To evaluate the effectiveness of our approach in each scenario, we conducted experiments using image-based and video-based classification. In the image-based classification setting, we treated each video frame as an individual presentation attempt and performed the classification on each frame independently. For video-based classification, we considered each video as a single presentation attempt. We aggregated the predictions from the first 'n' frames to obtain the final label and employed a majority vote mechanism. In the case where 'n' is an even number, the vote was biased towards the spoof label. The results of our study are presented in Table 6, showcasing the performance of our approach in both image-based and video-based classification scenarios.

When considering the image-based classification approach, our findings indicate that the GAN3 approach exhibited the highest performance in SiW protocol two but the lowest in protocol 3. Conversely, the KFGAN3 approach demonstrated the highest performance in protocol three but the lowest in protocol 2. Upon further investigation, we noticed a significant difference in the FID values between the KFGAN3 and GAN3 models, as displayed in Table 3. FID is a similarity measure between the generated and original samples, indicating that the KFGAN3-generated candidates exhibit more significant variability within the training set than the GAN3-generated samples. Furthermore, performing data augmentation after the split can contribute to an increase in the training set variability. The observed high variability appears to benefit the detection of unseen presentation attacks (protocol 3), while low variability appears beneficial for identifying unseen spoof mediums of the same type (protocol 2). These observations align with our earlier findings on the impact of data augmentation on sample variability, where a greater variability among spoof samples in the training set facilitates the identification of similarities, thus reducing both inter-set and intra-set variability.

In terms of the optimal frame window, our findings indicate that for protocol 2, data augmentation is unnecessary when using frame sizes of 5, 7, 10, and 15. Additionally, the baseline, traditional, and KFGAN3 approaches achieved state-of-the-art performance across both protocols when utilizing a frame size of 15. Notably, the KFGAN3 approach achieved comparable performance with reduced processing time using a window size of 10.

As the window size increases, we observed a trend in image-based classification with reduced error values. Moving on to the CASIA-FASD dataset, we found that the KFGAN3 approach outperformed the other approaches for both image and video-based classification when using a window size of 7. Although the other approaches achieved the same ACER and EER values with the same window size, the KFGAN3 approach consistently achieved the lowest values.

Based on our study, the optimal window size for video-based classification lies between the first 7 frames for CASIA-FASD and 15 frames for SiW. These findings suggest that the initial few video frames can effectively reveal whether a presentation is genuine or a spoof.

The Optimal Data Augmentation Percentage (DAP): Figure 10 demonstrates the optimal data augmentation percentages (DAP) for each approach for protocols 2 and 3. The GAN3 approach performs best with a DAP of 20% for protocol 2 and 0% for protocol 3. Similarly, the KFGAN3 approach achieves optimal results with a DAP of 20% for protocol 2 and 5% for protocol 3. On the other hand, the traditional approach shows optimal performance with a DAP of 5% for protocol 2 and 30% for protocol 3. To further support these findings, Tables 7 and 8 confirm that going beyond a DAP of 30% leads to diminishing returns in terms of equal error rate (EER) and average classification error rate (ACER) on the CASIA-FASD dataset. This suggests that increasing the data augmentation percentage beyond a certain threshold does not significantly improve the performance of the approaches. These results indicate the importance of selecting the appropriate data augmentation percentage for each protocol and approach, as exceeding or undershooting the optimal range can negatively impact performance.

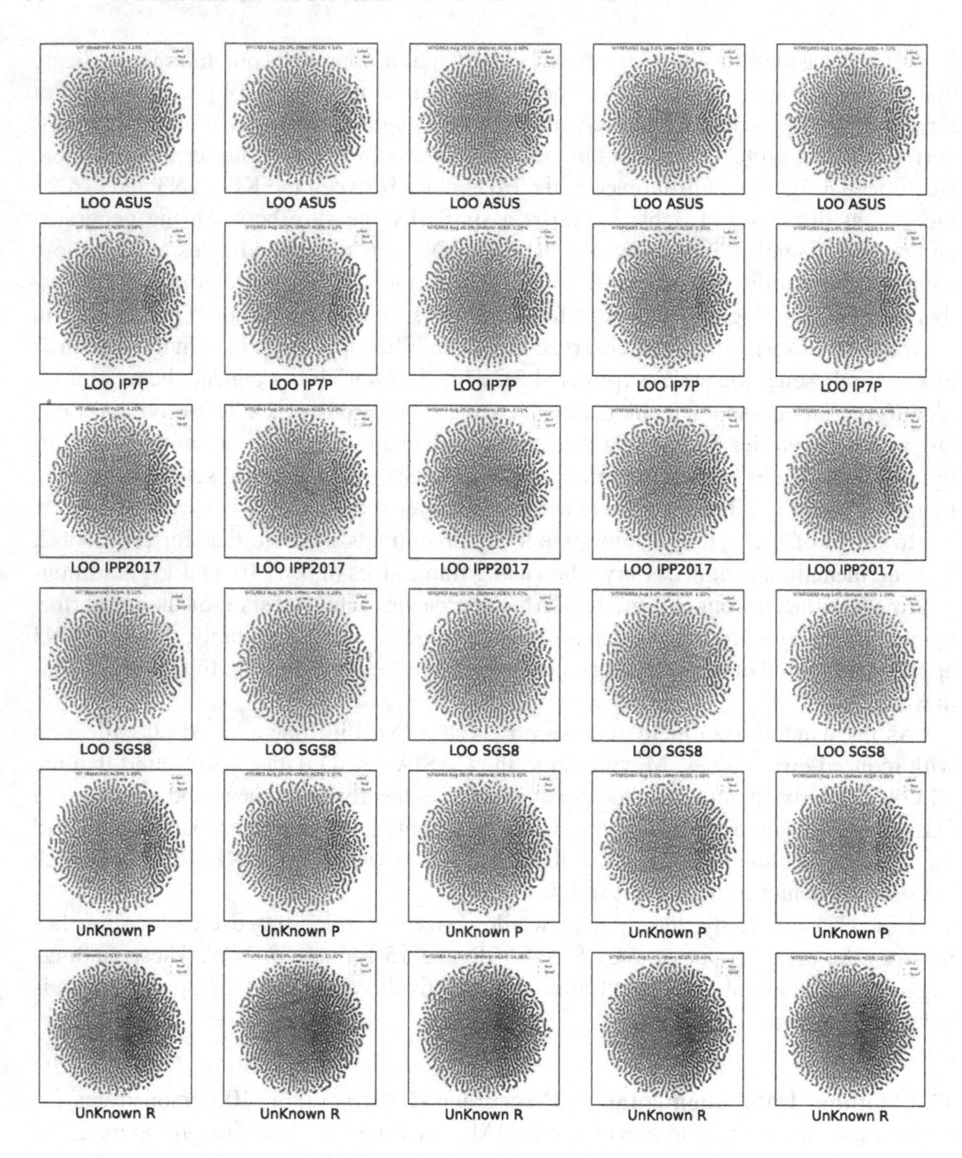

Fig. 11. A t-SNE plot for the top generative models evaluated on SiW protocols 2 and 3.

In our study, we analysed the performance of our top generative models on each SiW protocol using t-SNE plots, as depicted in Fig. 11. The analysis revealed exciting insights regarding the impact of data augmentation. We observed that data augmentation had a detrimental effect on the leave-one-out (LOO) SGS8 case of protocol 2. This suggests that the augmentation introduced noise or inconsistencies that hindered the model's performance for this specific case. Additionally, the LOO SGS8 and ASUS cases were identified as the primary factors contributing to the sub-par performance of the ViTKFGAN3 model in protocol 2. However, the ViTKFGAN3 model excelled in

Table 7. The performance of the baseline (ViT), traditional data augmentation (ViTT) and generative data augmentation (ViTGAN3, ViTKFGAN3) models for CASIA-FASD protocol 7 in terms of EER (%) for the image-based and video-based (window size of 7) classification approaches [33].

Model	Aug. (%)	Image EER (%)	Video EER (%)
ViT (Baseline)	0	1.75	2.01
ViTT (Before)	5	1.96	2.18
	10	2	1.83
	20	2.12	2.36
	30	2.18	2.36
ViTT (After)	5	1.89	1.65
	10	1.96	1.82
	20	2.13	2
	30	2.21	1.82
ViTGAN3 (Before)	5	1.77	1.65
	10	1.72	1.47
	20	1.75	1.71
	30	1.81	**1.29**
ViTGAN3 (After)	5	1.72	1.53
	10	1.8	1.83
	20	1.83	1.65
	30	1.87	1.83
ViTKFGAN3 (Before)	5	**1.71**	1.83
	10	**1.71**	**1.29**
	20	1.73	1.47
	30	1.85	1.65
ViTKFGAN3 (After)	5	1.78	1.83
	10	1.82	1.83
	20	1.86	2.01
	30	1.9	2.01

handling unknown attacks in protocol 3. Conversely, the ViTGAN3 model performed well in most cases, except for the LOO SGS8 case in protocol two, but struggled in protocol 3. Overall, we observed that a consistent data augmentation percentage throughout the protocols yielded good separation between real and spoof classes in the t-SNE plots, leading to generally satisfactory performance. However, employing a dynamic data augmentation percentage based on the specific characteristics of each case could further improve performance.

Table 8. Performance of the baseline (ViT), traditional data augmentation (ViTT), and generative data augmentation (ViTGAN3, ViTKFGAN3) models for CASIA-FASD protocol 7 in terms of ACER (%) for the image-based and video-based (window size of 7) classification approaches [33].

Model	Aug. (%)	Image ACER (%)	Video ACER (%)
ViT (Baseline)	0	1.42	1.42
ViTT (Before)	5	1.39	1.45
	10	1.38	1.27
	20	1.38	1.39
	30	1.41	1.45
ViTT (After)	5	1.37	**1.11**
	10	1.39	1.17
	20	1.41	1.2
	30	1.42	1.2
ViTGAN3 (Before)	5	1.37	1.2
	10	1.36	**1.11**
	20	1.41	1.36
	30	1.42	1.14
ViTGAN3 (After)	5	1.36	**1.11**
	10	1.35	1.27
	20	1.35	1.23
	30	1.36	1.3
ViTKFGAN3 (Before)	5	1.42	1.3
	10	1.36	**1.11**
	20	**1.34**	1.14
	30	1.38	1.23
ViTKFGAN3 (After)	5	1.36	1.27
	10	1.36	1.2
	20	1.35	1.36
	30	**1.34**	1.3

Benchmark Analysis: In the case of the SiW dataset, our independent frame analysis approach improved the performance of the baseline model. However, it fell short of competing with the current state-of-the-art independent methods in protocols 2 [9,31,42] and 3 [53]. These methods have achieved higher performance in terms of ACER. Nevertheless, our dependent frame analysis approach surpassed the current state-of-the-art dependent analysis methods for protocols 2 [46] and 3 [52], achieving a remarkable 0% ACER. Regarding the CASIA-FASD dataset, our approaches demonstrated competitive performance in protocol 7. However, they did not surpass the results achieved by [45] in terms of Equal Error Rate (EER).

Table 9. A benchmark of our generative data augmentation approach on SiW protocols 2 and 3.

Model	Classification Approach	Metric (%)	Protocol 2	Protocol 3
Auxiliary [30]	Video-based Analysis	APCER	0.57 ± 0.69	8.31 ± 3.81
		BPCER	0.57 ± 0.69	8.31 ± 3.8
		ACER	0.57 ± 0.69	8.31 ± 3.81
FAS-TD [44]	Video-based Analysis	APCER	0.08 ± 0.17	3.1 ± 0.79
		BPCER	0.21 ± 0.16	3.09 ± 0.83
		ACER	0.15 ± 0.14	3.1 ± 0.81
STASN [49]	Video-based Analysis	APCER	–	–
		ACER	–	–
		BPCER	0.15 ± 0.05	5.85 ± 0.85
BCN [50]	Image-based Analysis	APCER	0.08 ± 0.17	2.55 ± 0.89
		BPCER	0.15 ± 0	2.34 ± 0.47
		ACER	0.11 ± 0.08	2.45 ± 0.68
Disentangle [55]	Image-based Analysis	APCER	0.08 ± 0.17	9.35 ± 6.14
		BPCER	0.13 ± 0.09	1.84 ± 2.6
		ACER	0.1 ± 0.04	5.59 ± 4.37
CDCN [53]	Image-based Analysis	APCER	0 ± 0	1.67 ± 0.11
		BPCER	0.13 ± 0.09	1.76 ± 0.12
		ACER	0.06 ± 0.04	1.71 ± 0.11
CDCN++ [53]	Image-based Analysis	APCER	0 ± 0	1.97 ± 0.33
		BPCER	0.09 ± 0.1	1.77 ± 0.1
		ACER	0.04 ± 0.05	1.9 ± 0.15
NAS-FAS [52]	Image-based Analysis	APCER	0 ± 0	1.58 ± 0.23
		BPCER	0.09 ± 0.1	1.46 ± 0.08
		ACER	0.04 ± 0.05	1.52 ± 0.13
FAS-SGTD [43]	Video-based Analysis	APCER	0 ± 0	2.63 ± 3.72
		ACER	0.04 ± 0.08	2.92 ± 3.42
		ACER	0.02 ± 0.04	2.78 ± 3.57
STDN [31]	Image-based Analysis	APCER	0 ± 0	8.3 ± 3.3
		ACER	0 ± 0	7.5 ± 3.3
		BPCER	**0 ± 0**	7.9 ± 3.3
DRL-FAS [9]	Image-based Analysis	APCER	–	–
		BPCER	–	–
		ACER	0 ± 0	4.51 ± 0
FasTCo [46]	Video-based Analysis	APCER	0.02 ± 0.02	2.73 ± 0.91
		ACER	0 ± 0	1.28 ± 0.21
		BPCER	0.01 ± 0.01	2 ± 0.56
PatchNet [42]	Image-based Analysis	APCER	0 ± 0	3.06 ± 1.1
		BPCER	**0 ± 0**	1.83 ± 0.83
		ACER	0 ± 0	2.45 ± 0.45
ViTKFGAN3 (5%) After Split	Image-based Analysis	APCER	5.51 ± 5.88	13.79 ± 11.62
		BPCER	2.42 ± 2.77	0.95 ± 0.84
		ACER	3.97 ± 2.62	7.37 ± 5.45
ViTKFGAN3 (30%) After Split Window size of 15	Video-based Analysis	APCER	0.00 ± 0.00	0.00 ± 0.00
		BPCER	0.00 ± 0.00	0.00 ± 0.00
		ACER	**0.00 ± 0.00**	**0.00 ± 0.00**

Table 10. The metrics achieved by the approaches evaluated on CASIA-FASD protocol 7. *W-N* denotes the first *N* frames used in the Video-based analysis approach, and *-A* refers to data augmentation after the split.

Approach	Analysis Approach	EER (%)	ACER (%)
DoG [56]	Image-based Analysis	17.00	–
LBP [12]	Image-based Analysis	–	18.17
LBP [34]	Video-based Analysis	10.00	23.75
CNN [48]	Image-based Analysis	4.64	4.95
LBP [39]	Video-based Analysis	–	21.75
LBP [6]	Image-based Analysis	6.20	–
LSTM-CNN [47]	Video-based Analysis	5.17	5.93
LBP [7]	Image-based Analysis	3.20	–
Partial CNN [26]	Image-based Analysis	4.5	–
SURF [8]	Image-based Analysis	2.8	–
Patch-Depth CNN [2]	Image-based Analysis	2.67	2.27
Deep LBP [27]	Image-based Analysis	2.30	–
Hybrid CNN [28]	Image-based Analysis	2.2	–
Attention CNN [11]	Video-based Analysis	3.145	–
DropBlock [45]	Image-based Analysis	**1.12**	–
ViTGAN3-A (5%)	Image-based Analysis	1.72	1.36
ViTGAN3-A W-7 (5%)	Video-based Analysis	1.53	**1.11**
ViTKFGAN3-A W-7 (10%)	Video-based Analysis	1.83	**1.11**

Nonetheless, our dependent analysis approach achieved the best Average Classification Error Rate (ACER) among the methods that reported this metric, establishing it as the state-of-the-art method in terms of ACER for protocol 7. These findings highlight the efficacy of generative data augmentation in improving face antispoofing performance on the CASIA-FASD dataset. Overall, our results demonstrate that generative data augmentation is a promising approach for enhancing the performance of face antispoofing systems. While it may not outperform all state-of-the-art methods in every protocol or dataset, it offers competitive performance. It can be a valuable tool in combating face spoofing attacks.

6 Conclusion

In this study, we focused on optimising the performance of a vision transformer by leveraging generative and traditional data augmentation approaches. We trained StyleGAN3 models for each attack vector, generating candidate samples to enhance the training process. Additionally, we extended this approach by training StyleGAN3 models using keyframes instead of using all the frames in the training set (Tables 9 and 10).

By comparing the generated samples, we observed that using keyframes increased the variability among the training and validation sets while using all the frames increased the similarity. We found that higher variability was beneficial for detecting unknown presentation attacks, whereas higher similarity was advantageous for detecting unknown attacks of the same type. We explored the effectiveness of our approach in face antispoofing using image-based and video-based classification methods. Our findings indicated that the initial frames in a video were more effective in detecting face spoofing attacks than treating each frame independently. Specifically, employing keyframe data augmentation with the first 15 frames resulted in the best performance for protocols 2 and 3 of the Spoof in the Wild (SiW) dataset. Lastly, we observed that a consistent data augmentation percentage throughout the protocols yielded good separation between real and spoof classes in the t-SNE plots, but we suspect a dynamic data augmentation percentage could produce better results.

The results from SiW and CASIA-FASD datasets highlighted the efficacy of keyframe data augmentation as the most effective approach. Furthermore, we anticipate that augmenting training sets with generated spoof images can enhance the robustness of deep learning models against DeepFake attacks. Future work will investigate this aspect further and explore more advanced techniques in GAN image generation.

Acknowledgements. The support and resources from the South African Lengau cluster at the Centre for High-Performance Computing (CHPC) are gratefully acknowledged.

References

1. Ali, A., Deravi, F., Hoque, S.: Directional sensitivity of gaze-collinearity features in liveness detection. In: 2013 Fourth International Conference on Emerging Security Technologies, pp. 8–11 (2013). https://doi.org/10.1109/EST.2013.7
2. Atoum, Y., Liu, Y., Jourabloo, A., Liu, X.: Face anti-spoofing using patch and depth-based CNNs. In: 2017 IEEE International Joint Conference on Biometrics (IJCB), pp. 319–328 (2017). https://doi.org/10.1109/BTAS.2017.8272713
3. Bao, W., Li, H., Li, N., Jiang, W.: A liveness detection method for face recognition based on optical flow field. In: 2009 International Conference on Image Analysis and Signal Processing, pp. 233–236 (2009). https://doi.org/10.1109/IASP.2009.5054589
4. Ben Mabrouk, A., Zagrouba, E.: Abnormal behavior recognition for intelligent video surveillance systems: a review. Expert Syst. Appl. **91**, 480–491 (2018). https://doi.org/10.1016/j.eswa.2017.09.029, https://www.sciencedirect.com/science/article/pii/S0957417417306334
5. Bischoff, P.: Facial recognition technology (FRT): 100 countries analyzed - comparitech (2021). https://www.comparitech.com/blog/vpn-privacy/facial-recognition-statistics/
6. Boulkenafet, Z., Komulainen, J., Hadid, A.: Face anti-spoofing based on color texture analysis. In: 2015 IEEE International Conference on Image Processing (ICIP), pp. 2636–2640 (2015). https://doi.org/10.1109/ICIP.2015.7351280
7. Boulkenafet, Z., Komulainen, J., Hadid, A.: Face spoofing detection using colour texture analysis. IEEE Trans. Inf. Forensics Secur. **11**(8), 1818–1830 (2016). https://doi.org/10.1109/TIFS.2016.2555286
8. Boulkenafet, Z., Komulainen, J., Hadid, A.: Face antispoofing using speeded-up robust features and fisher vector encoding. IEEE Signal Process. Lett. **24**(2), 141–145 (2017). https://doi.org/10.1109/LSP.2016.2630740

9. Cai, R., Li, H., Wang, S., Chen, C., Kot, A.C.: DRL-FAS: a novel framework based on deep reinforcement learning for face anti-spoofing. IEEE Trans. Inf. Forensics Secur. **16**, 937–951 (2020)
10. Cao, Q., Shen, L., Xie, W., Parkhi, O.M., Zisserman, A.: VGGFace2: a dataset for recognising faces across pose and age. In: 2018 13th IEEE International Conference on Automatic Face & Gesture Recognition (FG 2018), pp. 67–74 (2018). https://doi.org/10.1109/FG.2018.00020
11. Chen, H., Hu, G., Lei, Z., Chen, Y., Robertson, N.M., Li, S.Z.: Attention-based two-stream convolutional networks for face spoofing detection. IEEE Trans. Inf. Forensics Secur. **15**, 578–593 (2020). https://doi.org/10.1109/TIFS.2019.2922241
12. Chingovska, I., Anjos, A., Marcel, S.: On the effectiveness of local binary patterns in face anti-spoofing. In: 2012 BIOSIG - Proceedings of the International Conference of Biometrics Special Interest Group (BIOSIG), pp. 1–7 (2012)
13. Daniel, N., Anitha, A.: A study on recent trends in face spoofing detection techniques. In: 2018 3rd International Conference on Inventive Computation Technologies (ICICT), pp. 583–586 (2018). https://doi.org/10.1109/ICICT43934.2018.9034361
14. George, A., Marcel, S.: On the effectiveness of vision transformers for zero-shot face anti-spoofing. In: 2021 IEEE International Joint Conference on Biometrics (IJCB), pp. 1–8 (2021). https://doi.org/10.1109/IJCB52358.2021.9484333
15. van der Haar, D.T.: Face antispoofing using shearlets: an empirical study. SAIEE Afr. Res. J. **110**(2), 94–103 (2019). https://doi.org/10.23919/SAIEE.2019.8732799
16. Hernandez-Ortega, J., Fierrez, J., Morales, A., Galbally, J.: Introduction to presentation attack detection in face biometrics and recent advances (2021). https://doi.org/10.48550/ARXIV.2111.11794, https://arxiv.org/abs/2111.11794
17. Heusel, M., Ramsauer, H., Unterthiner, T., Nessler, B., Hochreiter, S.: GANs trained by a two time-scale update rule converge to a local nash equilibrium (2017). https://doi.org/10.48550/ARXIV.1706.08500, https://arxiv.org/abs/1706.08500
18. ISO/IEC: International Organization for Standardization. Information technology - Biometric presentation attack detection - part 3: Testing and reporting. Technical report ISO/IEC FDIS 30107-3:2017(E), Geneva, CH (2017)
19. Jain, A.K., Flynn, P., Ross, A.A.: Handbook of Biometrics. Springer, New York (2007). https://doi.org/10.1007/978-0-387-71041-9
20. Johnson, J., Douze, M., Jégou, H.: Billion-scale similarity search with GPUs. IEEE Trans. Big Data **7**(3), 535–547 (2019)
21. Karras, T., Aittala, M., Hellsten, J., Laine, S., Lehtinen, J., Aila, T.: Training generative adversarial networks with limited data. In: Proceedings of the NeurIPS (2020)
22. Kingma, D.P., Ba, J.: Adam: a method for stochastic optimization (2014). https://doi.org/10.48550/ARXIV.1412.6980, https://arxiv.org/abs/1412.6980
23. Kolesnikov, A., et al.: An image is worth 16×16 words: transformers for image recognition at scale (2021)
24. Krishnan, K.S., Krishnan, K.S.: Vision transformer based COVID-19 detection using chest X-rays. In: 2021 6th International Conference on Signal Processing, Computing and Control (ISPCC), pp. 644–648 (2021). https://doi.org/10.1109/ISPCC53510.2021.9609375
25. Li, J.W.: Eye blink detection based on multiple Gabor response waves. In: 2008 International Conference on Machine Learning and Cybernetics, vol. 5, pp. 2852–2856 (2008). https://doi.org/10.1109/ICMLC.2008.4620894
26. Li, L., Feng, X., Boulkenafet, Z., Xia, Z., Li, M., Hadid, A.: An original face anti-spoofing approach using partial convolutional neural network. In: 2016 Sixth International Conference on Image Processing Theory, Tools and Applications (IPTA), pp. 1–6 (2016). https://doi.org/10.1109/IPTA.2016.7821013

27. Li, L., Feng, X., Jiang, X., Xia, Z., Hadid, A.: Face anti-spoofing via deep local binary patterns. In: 2017 IEEE International Conference on Image Processing (ICIP), pp. 101–105 (2017). https://doi.org/10.1109/ICIP.2017.8296251
28. Li, L., Xia, Z., Li, L., Jiang, X., Feng, X., Roli, F.: Face anti-spoofing via hybrid convolutional neural network. In: 2017 International Conference on the Frontiers and Advances in Data Science (FADS), pp. 120–124 (2017). https://doi.org/10.1109/FADS.2017.8253209
29. Liaw, R., Liang, E., Nishihara, R., Moritz, P., Gonzalez, J.E., Stoica, I.: Tune: a research platform for distributed model selection and training. arXiv preprint arXiv:1807.05118 (2018)
30. Liu, Y., Jourabloo, A., Liu, X.: Learning deep models for face anti-spoofing: Binary or auxiliary supervision. In: 2018 IEEE/CVF Conference on Computer Vision and Pattern Recognition, pp. 389–398 (2018). https://doi.org/10.1109/CVPR.2018.00048
31. Liu, Y., Stehouwer, J., Liu, X.: On disentangling spoof trace for generic face anti-spoofing. In: Vedaldi, A., Bischof, H., Brox, T., Frahm, J.-M. (eds.) ECCV 2020. LNCS, vol. 12363, pp. 406–422. Springer, Cham (2020). https://doi.org/10.1007/978-3-030-58523-5_24
32. Nagpal, C., Dubey, S.R.: A performance evaluation of convolutional neural networks for face anti spoofing. In: 2019 International Joint Conference on Neural Networks (IJCNN), pp. 1–8 (2019). https://doi.org/10.1109/IJCNN.2019.8852422
33. Orfao., J., van der Haar., D.: Keyframe and GAN-based data augmentation for face anti-spoofing. In: Proceedings of the 12th International Conference on Pattern Recognition Applications and Methods - ICPRAM, pp. 629–640. INSTICC, SciTePress (2023). https://doi.org/10.5220/0011648400003411
34. Pereira, T.D.F., et al.: Face liveness detection using dynamic texture. EURASIP J. Image video Process. **2014**(1), 1–15 (2014)
35. Pérez-Cabo, D., Jiménez-Cabello, D., Costa-Pazo, A., López-Sastre, R.J.: Deep anomaly detection for generalized face anti-spoofing. In: 2019 IEEE/CVF Conference on Computer Vision and Pattern Recognition Workshops (CVPRW), pp. 1591–1600 (2019). https://doi.org/10.1109/CVPRW.2019.00201
36. Rodríguez, J., Lozano, J.: Repeated stratified k-fold cross-validation on supervised classification with naive bayes classifier: an empirical analysis (2007)
37. Shahapure, K.R., Nicholas, C.: Cluster quality analysis using silhouette score. In: 2020 IEEE 7th International Conference on Data Science and Advanced Analytics (DSAA), pp. 747–748 (2020). https://doi.org/10.1109/DSAA49011.2020.00096
38. Shorten, C., Khoshgoftaar, T.: A survey on image data augmentation for deep learning. J. Big Data **6** (2019). https://doi.org/10.1186/s40537-019-0197-0
39. Tirunagari, S., Poh, N., Windridge, D., Iorliam, A., Suki, N., Ho, A.T.S.: Detection of face spoofing using visual dynamics. IEEE Trans. Inf. Forensics Secur. **10**(4), 762–777 (2015). https://doi.org/10.1109/TIFS.2015.2406533
40. Ur Rehman, Y.A., Po, L.M., Liu, M.: Deep learning for face anti-spoofing: an end-to-end approach. In: 2017 Signal Processing: Algorithms, Architectures, Arrangements, and Applications (SPA), pp. 195–200 (2017). https://doi.org/10.23919/SPA.2017.8166863
41. O'Mahony, N., et al.: Deep learning vs. traditional computer vision. In: Arai, K., Kapoor, S. (eds.) CVC 2019. AISC, vol. 943, pp. 128–144. Springer, Cham (2020). https://doi.org/10.1007/978-3-030-17795-9_10
42. Wang, C.Y., Lu, Y.D., Yang, S.T., Lai, S.H.: PatchNet: a simple face anti-spoofing framework via fine-grained patch recognition. In: 2022 IEEE/CVF Conference on Computer Vision and Pattern Recognition (CVPR), pp. 20249–20258 (2022). https://doi.org/10.1109/CVPR52688.2022.01964
43. Wang, Z., et al.: Deep spatial gradient and temporal depth learning for face anti-spoofing. In: 2020 IEEE/CVF Conference on Computer Vision and Pattern Recognition (CVPR), pp. 5041–5050 (2020). https://doi.org/10.1109/CVPR42600.2020.00509

44. Wang, Z., et al.: Exploiting temporal and depth information for multi-frame face anti-spoofing. arXiv preprint arXiv:1811.05118 (2018)
45. Wu, G., Zhou, Z., Guo, Z.: A robust method with dropblock for face anti-spoofing. In: 2021 International Joint Conference on Neural Networks (IJCNN), pp. 1–8 (2021). https://doi.org/10.1109/IJCNN52387.2021.9533595
46. Xu, X., Xiong, Y., Xia, W.: On improving temporal consistency for online face liveness detection system. In: 2021 IEEE/CVF International Conference on Computer Vision Workshops (ICCVW), pp. 824–833 (2021). https://doi.org/10.1109/ICCVW54120.2021.00097
47. Xu, Z., Li, S., Deng, W.: Learning temporal features using LSTM-CNN architecture for face anti-spoofing. In: 2015 3rd IAPR Asian Conference on Pattern Recognition (ACPR), pp. 141–145 (2015). https://doi.org/10.1109/ACPR.2015.7486482
48. Yang, J., Lei, Z., Li, S.Z.: Learn convolutional neural network for face anti-spoofing. arXiv preprint arXiv:1408.5601 (2014)
49. Yang, X., et al.: Face anti-spoofing: model matters, so does data. In: 2019 IEEE/CVF Conference on Computer Vision and Pattern Recognition (CVPR), pp. 3502–3511 (2019). https://doi.org/10.1109/CVPR.2019.00362
50. Yu, Z., Li, X., Niu, X., Shi, J., Zhao, G.: Face anti-spoofing with human material perception. In: Vedaldi, A., Bischof, H., Brox, T., Frahm, J.-M. (eds.) ECCV 2020. LNCS, vol. 12352, pp. 557–575. Springer, Cham (2020). https://doi.org/10.1007/978-3-030-58571-6_33
51. Yu, Z., Qin, Y., Li, X., Zhao, C., Lei, Z., Zhao, G.: Deep learning for face anti-spoofing: a survey (2021). https://doi.org/10.48550/ARXIV.2106.14948, https://arxiv.org/abs/2106.14948
52. Yu, Z., Wan, J., Qin, Y., Li, X., Li, S.Z., Zhao, G.: NAS-FAS: static-dynamic central difference network search for face anti-spoofing. IEEE Trans. Pattern Anal. Mach. Intell. **43**(9), 3005–3023 (2020)
53. Yu, Z., et al.: Searching central difference convolutional networks for face anti-spoofing. In: 2020 IEEE/CVF Conference on Computer Vision and Pattern Recognition (CVPR), pp. 5294–5304 (2020). https://doi.org/10.1109/CVPR42600.2020.00534
54. Zhang, K., Zhang, Z., Li, Z., Qiao, Y.: Joint face detection and alignment using multitask cascaded convolutional networks. IEEE Signal Process. Lett. **23**(10), 1499–1503 (2016). https://doi.org/10.1109/LSP.2016.2603342
55. Zhang, K.-Y., et al.: Face anti-spoofing via disentangled representation learning. In: Vedaldi, A., Bischof, H., Brox, T., Frahm, J.-M. (eds.) ECCV 2020. LNCS, vol. 12364, pp. 641–657. Springer, Cham (2020). https://doi.org/10.1007/978-3-030-58529-7_38
56. Zhang, Z., Yan, J., Liu, S., Lei, Z., Yi, D., Li, S.Z.: A face antispoofing database with diverse attacks. In: 2012 5th IAPR International Conference on Biometrics (ICB), pp. 26–31 (2012). https://doi.org/10.1109/ICB.2012.6199754

Improving Person Re-identification Through Low-Light Image Enhancement

Oliverio J. Santana(✉), Javier Lorenzo-Navarro, David Freire-Obregón, Daniel Hernández-Sosa, and Modesto Castrillón-Santana

SIANI, Universidad de Las Palmas de Gran Canaria, Las Palmas de Gran Canaria, Spain
oliverio.santana@ulpgc.es

Abstract. Person re-identification (ReID) is a popular area of research in the field of computer vision. Despite the significant advancements achieved in recent years, most of the current methods rely on datasets containing subjects captured with good lighting under static conditions. ReID presents a significant challenge in real-world sporting scenarios, such as long-distance races that take place over varying lighting conditions, ranging from bright daylight to night-time. Unfortunately, increasing the exposure time on the capture devices to mitigate low-light environments is not a feasible solution, as it would result in blurry images due to the motion of the runners. This paper surveys several low-light image enhancement methods and finds that including an image pre-processing step in the ReID pipeline before extracting the distinctive body features of the subjects can lead to significant improvements in performance.

Keywords: Computer vision · Person re-identification · Sporting event · Ultra-trail running · In the wild dataset · Low-light image enhancement

1 Introduction

Effortlessly identifying all individuals in an image relies on a thorough semantic comprehension of the people in the scene. However, despite it is a natural capability for humans, modern visual recognition systems still struggle to accomplish this task, and thus the research community is devoting much effort to this topic. Person re-identification (ReID) is an important goal in this field, which involves matching individuals captured at various locations and times.

In recent years, significant ReID advancements have been made [23,25]. The evolution of image capture devices has led to new, more challenging datasets that go beyond short-term ReID with homogeneous lighting conditions. Long-term ReID involves the challenge of handling substantial variability in space and time. For example, individuals captured in such scenarios may present pose variations and occlusions, multi-scale

This work is partially funded by the ULPGC under project ULPGC2018-08, by the Spanish Ministry of Science and Innovation under project PID2021-122402OB-C22, and by the ACIISI-Gobierno de Canarias and European FEDER funds under projects ProID2020010024, ProID2021010012, ULPGC Facilities Net, and Grant EIS 2021 04.

M. De Marsico et al. (Eds.): ICPRAM 2023, LNCS 14547, pp. 95–110, 2024.
https://doi.org/10.1007/978-3-031-54726-3_6

detections, appearance inconsistencies due to clothing changes, and various environmental and lighting shifts.

Current face recognition methods utilized in standard ReID situations have demonstrated poor performance, as highlighted by recent studies [2,3]. This is primarily attributed to the requirement for high-resolution facial images of good quality. Consequently, researchers has recently focused on low-light image enhancement techniques to address illumination variations in the target images.

Images that are not well illuminated tend to have low contrast and a high level of noise [26]. Adjusting the illumination properties of these images can introduce significant issues, such as over-saturation in regions with high-intensity pixels. As a result, the community has proposed various image enhancement techniques to tackle this problem.

The Retinex Theory [14], which focuses on the dynamic range and color constancy of the images and recovers contrast by precisely estimating illumination, is the foundation on which traditional low-light image enhancement methods are built. However, these methods may produce color distortion or introduce unnatural effects while enhancing image brightness [9]. Deep learning techniques have recently been developed in response, making significant strides in addressing this challenging task [35]. These techniques have progressed in two areas: improving image quality and reducing processing time.

Although there have been promising advances in low-light image enhancement, we feel that there is still much potential for improvement. For example, only a few recent studies have evaluated their techniques on video clips [16,21]. Unlike static images, video clips feature a continuous variation in lighting signals that can impact subsequent processes, such as person ReID.

Considering the current trend towards ReID in dynamic scenarios, we introduced a novel ReID pipeline in our previous work [29] that integrates software image enhancement methods and ReID. Both tasks are approached differently: the former utilizes information from the entire scene to improve the overall image quality, while the latter concentrates on a particular region of interest in the enhanced image.

We evaluated our proposal using various state-of-the-art image enhancement techniques, analyzing both traditional methods and deep learning approaches on a sporting dataset that includes 109 distinct identities in a 30-hour race under varying lighting conditions, settings, and accessories. Regarding our previous paper, we have complemented our results by assessing two different gamma correction factor values for the traditional algorithms. Additionally, we have included the evaluation of a new deep learning solution.

Our findings demonstrate the difficulty of performing runner ReID in these challenging conditions, especially when dealing with images captured in poor lighting. Our outcomes also reveal that deep learning techniques can provide better performance than traditional methods, as they take into account the overall context of the image rather than focusing solely on individual pixels. In this extended version of the paper, we provide an ablation study that considers the position of the image enhancer within the pipeline, illustrating why the enhancer should work with the entire image instead of processing only the region of interest.

In light of these results, we believe that our work is a valuable initial step towards combining different image processing methods to improve ReID performance under difficult conditions. Nonetheless, there is still considerable room for enhancing ReID performance in this scenario. We have expanded our original analysis with a detailed discussion of some cases in which the deep learning models made ReID mistakes and the reasons behind these errors, providing more insight on the inner processing of our pipeline.

The remainder of this paper is organized as follows. Section 2 reviews previous related work. Section 3 describes our ReID pipeline. Section 4 presents our dataset and evaluation results. Section 5 discusses these results and, finally, Sect. 6 presents our concluding remarks.

2 Related Work

Numerous techniques have been suggested to improve the quality of images taken in low-light situations. Typically, the Retinex Theory [14] serves as the basis for traditional illumination enhancement models, splitting images into two components: reflectance and illumination. While some approaches have been developed for improving images using both these components [6,12,15,17,31,34], they tend to require significant computational resources.

Guo et al. [7] introduced a technique for enhancing low-light images that simplifies the solution space by solely estimating the illumination component. To achieve this, an illumination map is generated by computing the maximum value of each RGB color channel for every pixel in the image. The map is then refined using an augmented Lagrangian algorithm that takes into account the illumination structure. Liu et al. [19] developed an efficient approach based on a membership function and gamma corrections, which reduces the computational cost compared to traditional methods and prevents image over-enhancement or under-enhancement.

It should be noted that these methods are specifically designed for low-light images and optimized to enhance underexposed images. On the other hand, Zhang et al. [36] proposed a dual-illumination-map estimation technique that generates two different maps for the original and inverted images. These maps can correct the underexposed and overexposed regions of the image.

The emergence of deep learning has led to the development of image enhancement techniques based on neural networks. Wei et al. [32] drew on the Retinex Theory to design Retinex-Net, a network that decomposes images in their two components and recomposes them after lighting correction. However, most deep learning approaches have been developed from scratch, without considering the Retinex Theory structural limitations [1,9,13,24,30]. Recently, Hao et al. [8] proposed a two-stage decoupled network that provides results comparable to or better than other current approaches. The first network learns the scene structure and illumination distribution to generate an image resembling optimal lighting conditions. The second network further enhances the image by reducing noise and color distortion.

Overall, various computer vision problems have previously faced the challenges of low-illumination images. Liu et al. [18] provided a comprehensive review of low-light enhancement techniques and demonstrated the results of a face detection task.

The authors also introduced the VE-LOL-L dataset, which was gathered under low-light conditions and annotated for face detection. Ma et al. [22] introduced the TMD^{2}L distance learning method based on Local Linear Models and triplet loss for video re-identification. The effectiveness of the method was compared to other low-illumination enhancement techniques on three datasets: two simulated using PRID 2011 and iLIDS-VID, and a newly compiled LIPS dataset. Unlike Ma's approach, which aims to obtain person descriptors directly from low-illumination images, our proposal is closer to Liu's, involving an image pre-processing stage but in a ReID context instead of face detection.

3 ReID Pipeline

We conducted our experiments over the TGC2020 dataset [25]. This dataset has been previously used to address various computer vision challenges, including facial expression recognition [27], bib number recognition [10], and action quality assessment [4,5]. The TGC2020 dataset comprises images of runners captured at various recording points (RPs), illustrated in Fig. 1.

Our primary objective is to explore the effectiveness of several enhancement techniques on footage captured in wild environments and evaluate these techniques according to ReID performance. However, in order to achieve accurate ReID results, it is essential to eliminate distractors from the scenes, such as other runners, staff, and spectators.

Our ReID pipeline, initially proposed in [29], has three primary components, as shown in Fig. 2. First, the input footage is processed through a low-light image enhancement method to improve the visibility of the runners, resulting in better-quality footage. We evaluated various techniques to enhance the footage, including LIME [7], the dual estimation method [36], Retinex-Net [32], and the decoupled low-light image enhancement method [8].

Once the improved footage has been obtained, our pipeline focuses on identifying the precise locations of the runners within the scene. We first utilize a body detector based on Faster R-CNN with Inception-V2 pre-trained with the COCO dataset [11] to detect the bodies in the footage. Then, we employ DeepSORT to track the runners in the scene [33]. This tracker utilizes a combination of deep descriptors to integrate appearance and position information obtained through the Kalman filter used in the SORT algorithm.

After cropping the runner bodies from the footage, we extract the embeddings of each runner using the AlignedReID++ model [20] trained on the Market1501 dataset [37]. With these embeddings, which serve as features, we evaluate the effectiveness of our ReID pipeline using the mean average precision score (mAP) as the performance metric. This metric is particularly suitable for datasets where the same individual may appear multiple times in the gallery, as it considers all occurrences of each runner. The mAP score is calculated using the well-known Eq. 1.

$$mAP = \frac{\sum_{i=1}^{k} AP_i}{k} \quad (1)$$

Fig. 1. Examples of images of the race leading runner(s) captured in each one of the recording points (RP1 to RP5). Figure from [29].

In this equation, AP_i refers to the area under the precision-recall curve of probe i, and k is the total number of runners in the probe. We experimented with both Cosine and Euclidean distances to compute the mAP scores and found that both methods produced similar results. Consequently, we only present the results obtained using the Euclidean distance.

4 Experiments and Results

The TGC2020 dataset [25] contains images of the Transgrancanaria (TGC) ultra-trail race participants. Specifically, the recordings were taken during the TGC Classic event held in 2020, in which participants were required to complete a 128-kilometer journey on foot within 30 hour. The winners devoted roughly 13 hour to finish the race, but most runners needed longer. Since the race began at 11 pm, the runners covered the first 8 hour during nighttime.

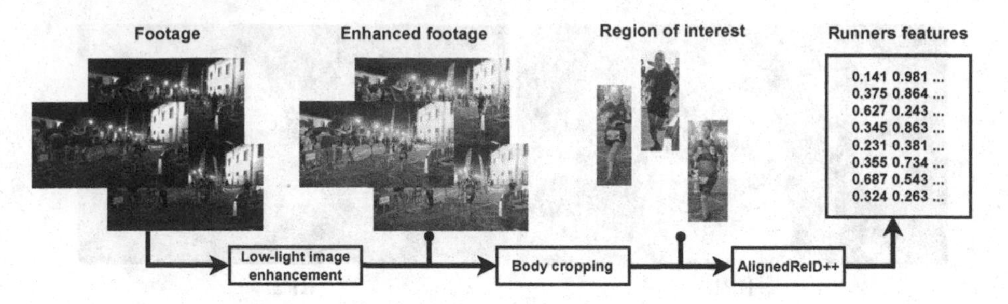

Fig. 2. Proposed ReID pipeline. The process comprises three main components: footage enhancement, region of interest cropping, and feature extraction. Figure from [29].

Table 1. Location of each RP and starting time of image recording. The first two RPs were recorded during the night, and thus, only artificial illumination was available for capturing images. Data from [29].

	location	km	time
RP1	Arucas	16.5	00:06
RP2	Teror	27.9	01:08
RP3	Presa de Hornos	84.2	07:50
RP4	Ayagaures	110.5	10:20
RP5	Parquesur	124.5	11:20

Table 1 provides information regarding the distance from the starting line to each recording point (RP) on the race track, as well as the time at which the images began to be recorded. The dataset consists of images taken at five different RPs: RP1 and RP2 were captured during the first hours of the race under nightlight conditions, while RP4 and RP5 were captured later in daylight conditions. RP3 images present an intermediate scenario, as they were taken during the early morning hours at a location where sunlight was obstructed by trees and mountains, resulting in a situation more akin to twilight than full sunlight. As expected, the lighting conditions of the images captured at each RP vary significantly. This circumstance is illustrated in Fig. 1.

4.1 ReID on the Original Dataset

To evaluate ReID performance, we analyze the images of the 109 runners identified in every RP. In addition, to maintain the temporal progression of the runners throughout the race, we only probe each runner against the galleries of the previous recording points. Table 2 shows the ReID mAP score using RP2, RP3, RP4, and RP5 as probes, each preceding RP serving as the gallery.

The best ReID results are achieved when RP5 is probed against RP4 as the gallery, since both RPs were recorded in full daylight. Conversely, the worst performance is observed when using RP3 as the probe, since it was captured in a peculiar intermediate lighting situation and has only low-light night images available for use as the

Table 2. ReID results (mAP) for the original dataset, i.e. without applying any of the pre-processing methods to the images. Data from [29].

	Gallery			
	RP1	RP2	RP3	RP4
RP2	14.80			
RP3	9.11	7.17		
RP4	22.66	13.13	19.21	
RP5	22.41	20.32	15.09	48.29

gallery. Overall, performance is relatively low, highlighting the significant challenge runner ReID presents in uncontrolled outdoor environments.

4.2 ReID on Footage Improved with Retinex Algorithms

Table 3 shows the results obtained using two algorithms based on the Retinex Theory: the low-light image enhancement method (LIME) proposed by Guo et al. [7] and the dual estimation method (Dual) proposed by Zhang et al. [36]. It is important to note that, to obtain an enhanced dataset, we apply the algorithm homogeneously to all the images in the original dataset. That is, we do not differentiate between nighttime and daytime images .

Table 3. ReID results (mAP) for the dataset enhanced using Retinex algorithms and two different values of the gamma correction factor. The number in brackets shows the variation compared to the original dataset.

	Gallery	LIME γ 0.3	LIME γ 0.6	Dual γ 0.3	Dual γ 0.6
RP2	RP1	21.55 (+6.8)	21.58 (+6.8)	19.95 (+5.2)	21.90 (+7.1)
RP3	RP1	13.89 (+4.8)	11.82 (+2.7)	9.72 (+0.6)	9.61 (+0.5)
	RP2	10.78 (+3.6)	11.73 (+4.6)	8.75 (+1.6)	9.44 (+2.3)
RP4	RP1	22.44 (−0.2)	19.69 (−3.0)	21.77 (−0.9)	16.05 (−6.6)
	RP2	16.98 (+3.9)	18.44 (+5.3)	16.55 (+3.4)	14.98 (+1.9)
	RP3	20.15 (+0.9)	20.35 (+1.1)	18.87 (−0.3)	17.75 (−1.5)
RP5	RP1	24.28 (+1.9)	20.69 (−1.7)	22.83 (+0.4)	18.11 (−4.3)
	RP2	26.34 (+6.0)	21.15 (+0.8)	24.51 (+4.2)	19.08 (−1.2)
	RP3	16.77 (+1.7)	15.71 (+0.6)	14.37 (−0.7)	13.89 (−1.2)
	RP4	48.01 (−0.3)	42.92 (−5.4)	44.75 (−3.5)	38.91 (−9.4)

A key configuration parameter for both algorithms is the gamma correction factor. As a general rule, previous works in the literature use values close to one, but they usually apply this correction factor to high-quality images. Due to the wild nature of our scenario, the images in the TGC2020 dataset have lower quality and suffer from motion blur. If the gamma value is increased excessively, the images become noisier and hinder ReID performance.

Applying the LIME algorithm noticeably improves the performance in the first three RPs, which are the most impacted by bad lighting conditions. However, the improvement is less apparent in the RPs that are already well illuminated and, in some cases, there may even be a slight performance degradation. Table 3 shows the results of the LIME algorithm application with two gamma values: 0.3 and 0.6. The higher value may provide better results in some nighttime RP combinations, but it can be very disruptive for the best-lit RPs, making it preferable to use the lower value.

On the other hand, the Dual algorithm yields inferior results compared to the LIME algorithm. Although both algorithms are based on the same mathematical principles, they are targeted at different goals. The LIME algorithm is designed to enhance the underexposed regions of the image, while the Dual algorithm aims to improve both the underexposed and the overexposed parts. These results indicate that overexposure is not a significant concern for runner ReID with this dataset. Therefore, in this particular scenario, opting for the algorithm that targets underexposed areas is advantageous.

4.3 ReID on Footage Improved with Deep Learning Approaches

In contrast to algorithmic methods, neural networks do not apply a series of preset operations through an iterative process. During their training phase, the networks learn the operations they have to apply to obtain the desired results. However, this does not necessarily mean that they will provide better performance. Table 4 shows the ReID performance on a dataset enhanced using the Retinex-Net model proposed by Wei et al. [32]. Overall, the performance obtained does not improve on the performance provided by the LIME algorithm. This indicates that the network design, intended to mimic the behavior of the Retinex algorithms, is not well suited for this problem. In order to

Table 4. ReID results (mAP) for the dataset enhanced using deep learning models. The number in brackets shows the variation compared to the original dataset.

	Gallery	Retinex-Net	Decoupled Net-I	Decoupled Net-II
RP2	RP1	15.54 (+0.7)	24.78 (+10.0)	26.99 (+12.2)
RP3	RP1	9.76 (+0.7)	19.17 (+10.1)	13.94 (+ 4.8)
	RP2	8.66 (+1.5)	18.14 (+11.0)	12.35 (+ 5.2)
RP4	RP1	16.05 (−6.6)	25.92 (+ 3.3)	23.52 (+ 0.9)
	RP2	15.53 (+2.4)	24.46 (+11.3)	23.61 (+10.5)
	RP3	18.43 (−0.8)	24.84 (+ 5.6)	22.82 (+ 3.6)
RP5	RP1	18.68 (−3.7)	27.00 (+ 4.6)	25.32 (+ 2.9)
	RP2	19.23 (−1.1)	29.22 (+ 8.9)	27.62 (+ 7.3)
	RP3	15.45 (+0.4)	19.59 (+ 4.5)	18.21 (+ 3.1)
	RP4	40.19 (−8.1)	56.41 (+ 8.1)	51.64 (+ 3.4)

take advantage of the full potential of deep learning, it would be advisable to design the neural network without including this kind of preconceived restriction.

The decoupled low-light image enhancement method proposed by Hao et al. [8] learns to interpret the general context of the images and provides the best ReID performance of all tested methods. Table 4 shows separately the results of applying only the first stage (Decoupled Net-I) and applying both stages (Decoupled Net-II). Interestingly, the best ReID results are obtained by applying only the first stage, which is responsible for enhancing image lighting. The color and noise corrections applied by the second stage only provide better ReID results in a single case.

4.4 Ablation Study: Position of the Enhancement Stage

Our original pipeline design [29] implements the image enhancement stage before the runner body cropper, allowing the enhancement methods to work with the whole image and exploit the general context of the scene. To gain a better understanding, we have evaluated the best approach, the decoupled low-light image enhancement method, while moving the enhancement stage after the body cropper. Hence the neural network would only be able to work with the region of the image where the runner is located.

Table 5. ReID results (mAP) for the dataset enhanced using the decoupled low-light image enhancement method after cropping the runner body images. The number in brackets shows the variation compared to the results obtained enhancing the images before cropping the runner bodies, as proposed in the original pipeline.

	Gallery	Decoupled Net-I	Decoupled Net-II
RP2	RP1	24.79 (+0.01)	27.18 (+0.19)
RP3	RP1	19.00 (−0.17)	13.17 (−0.77)
	RP2	18.29 (+0.15)	12.62 (+0.27)
RP4	RP1	25.72 (−0.20)	23.71 (+0.19)
	RP2	24.47 (+0.01)	25.21 (+1.60)
	RP3	24.88 (+0.04)	20.32 (−2.50)
RP5	RP1	26.52 (−0.48)	24.52 (−0.80)
	RP2	29.44 (+0.22)	28.98 (+1.36)
	RP3	19.47 (−0.12)	18.53 (+0.32)
	RP4	56.68 (+0.27)	47.35 (−4.29)

The results in Table 5 show that, in most cases, the impact of this change in the pipeline is negligible. Nevertheless, when the change leads to performance degradation, the variation in the results tends to be more significant than when it leads to performance improvement. Consequently, we consider that it is preferable to maintain the original pipeline design.

Interestingly, this change in the pipeline is more relevant when we apply the two stages of the decoupled low-light image enhancement method. This indicates that the general context of the image is more helpful in correcting color distortions and reducing noise than in improving scene lighting. In any case, the performance obtained applying only the first stage is still higher, making it the best choice among the evaluated methods.

5 Discussion

It is a well-established fact that computers do not perceive images in the same way as humans do. An enhanced image that appears the most appealing to human eyes may not be suitable for an automatic ReID system. Figure 3 demonstrates this by displaying the nighttime images of two runners in RP2. The images generated by the second stage of the decoupled low-light image enhancement method seem to be the best choice. However, the best ReID performance is achieved using the images generated by the first stage.

To provide a better insight into why this behavior occurs, Fig. 4 shows a runner in RP5 that is incorrectly identified in the RP2 gallery. The error is understandable because the clothing of the correct runner is very similar to the clothing of the runner incorrectly selected by the ReID pipeline. Even the white visor the probe runner wears in RP5 can be mistaken by the pipeline for the headlamp glow in the images of the two runners in RP2.

The main difference between the two runners is the color of the socks. When only the first stage of the decoupled low-light image enhancement method is applied, the enhanced image of the correct runner gets a shade of pink in the socks that is very similar to the one on the probe image, which makes it easier for the ReID pipeline to provide a correct identification. However, when the second stage is also applied, the color is modified to make the image more appealing, resulting in a pink shade that differs further from the original, leading the ReID pipeline to fail again.

Figure 5 shows another wrong identification example, this time of a probe runner in RP4 that is also incorrectly identified in the RP2 gallery. Once again, the clothing of both the correct runner and the runner incorrectly chosen by the pipeline are very similar. Pre-processing the images with the first stage of the decoupled low-light image enhancement method emphasizes the differences between the shade of blue of the shirts and, especially, the precise color of the backpacks, which in the case of the correct runner looks more violet than black. However, these hues are lost when the second stage adjusts the images, so the ReID fails again.

These results highlight the difficulty of choosing the best images for a ReID process based solely on human observation. It is improbable that a single method would be deemed optimal in all situations, which means that the method chosen should depend on the particular features of the pre-processed images. Finding the ideal approach for this purpose is a fascinating area for future research. It would benefit the community to establish a systematic method for selecting the most appropriate method for enhancing low-light images based on the various quality assessment metrics [35] suggested in the literature.

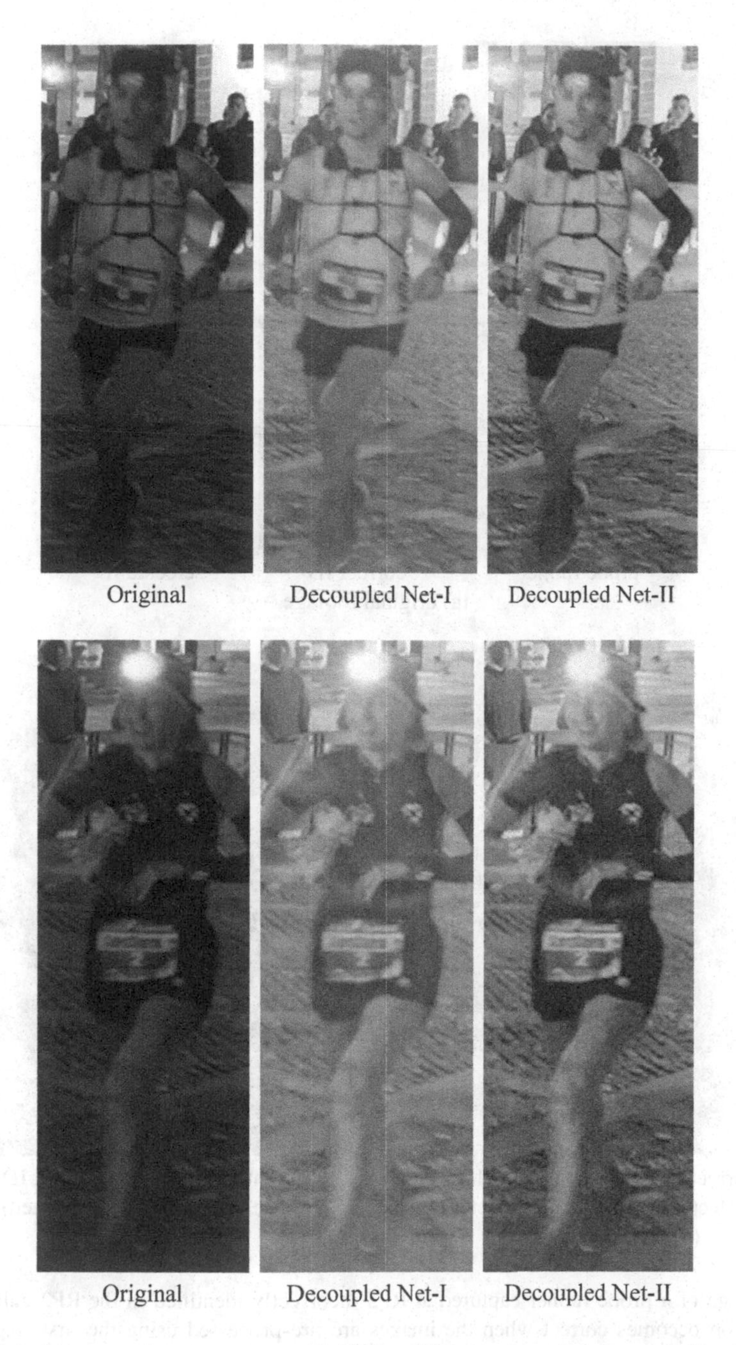

Fig. 3. Original images of runners captured in RP2 and improved versions generated by the decoupled low-light image enhancement method. Figure adapted from [29]. (Color figure online)

Fig. 4. Image of a probe runner captured at RP5 incorrectly identified in the RP2 gallery. The identification becomes correct when the images are pre-processed using the first stage of the decoupled low-light image enhancement method, but it becomes incorrect again when both stages are used. (Color figure online)

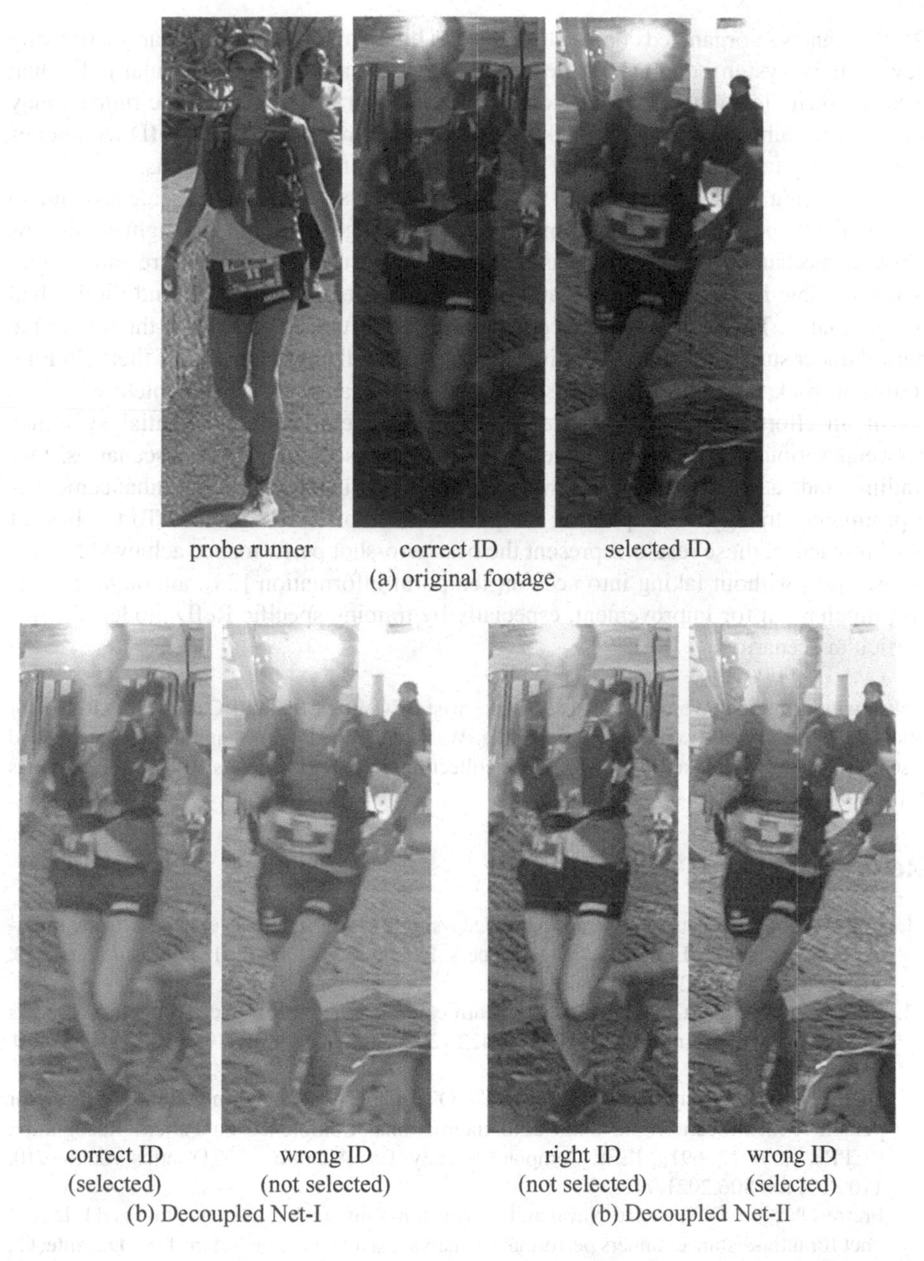

Fig. 5. Image of a probe runner captured at RP5 incorrectly identified in the RP4 gallery. The identification becomes correct when the images are pre-processed using the first stage of the decoupled low-light image enhancement method, but it becomes incorrect again when both stages are used. (Color figure online)

6 Conclusions

While runners in organized competitions wear a bib number and usually carry a tracking device, these systems do not provide an actual identification of the particular individual wearing them, leaving room for potential cheating. For instance, multiple runners may share a race bib to enhance the ranking of the bib owner. Therefore, ReID techniques are necessary to ensure the integrity of the results in running competitions.

Performing runner ReID in real-world environments is a highly intricate task due to the frequently poor lighting conditions. Improving image quality in low-light conditions through mechanical means, such as increasing exposure time to capture more light, is not feasible in this scenario because it would result in motion blur and diminished image quality. This problem is exacerbated in long-distance races, since the longer the period under study, the more likely it is that runners will alter elements of their clothing (outwear, backpack, glasses, bib position...) or even change clothes completely.

In an effort to overcome these challenges, we explore the potential synergies between various methods that were originally proposed for different scenarios. Our findings indicate that utilizing software-based techniques for lighting enhancement is a promising strategy for improving the performance of ReID systems. To the best of our knowledge, these results represent the best zero-shot performance achieved for this dataset [25] without taking into account temporal information [28], although there is still much room for improvement, especially by training specific ReID models for this particular scenario.

Acknowledgements. We would like to thank Arista Eventos SLU and Carlos Díaz Recio for granting us the use of Transgrancanaria media. We would also like to thank the volunteers and researchers who have taken part in the data collection and annotation, as well as the previous papers of this project.

References

1. Cai, J., Gu, S., Zhang, L.: Learning a deep single image contrast enhancer from multi-exposure images. IEEE Trans. Image Process. **27**(4), 2049–2062 (2018). https://doi.org/10.1109/TIP.2018.2794218
2. Cheng, Z., Zhu, X., Gong, S.: Face re-identification challenge: are face recognition models good enough? Pattern Recogn. **107**, 107422 (2020). https://doi.org/10.1016/j.patcog.2020.107422
3. Dietlmeier, J., Antony, J., McGuinness, K., O'Connor, N.E.: How important are faces for person re-identification? In: 2020 25th International Conference on Pattern Recognition (ICPR), pp. 6912–6919. IEEE Computer Society, Los Alamitos (2021). https://doi.org/10.1109/ICPR48806.2021.9412340
4. Freire-Obregón, D., Lorenzo-Navarro, J., Castrillón-Santana, M.: Decontextualized I3D convnet for ultra-distance runners performance analysis at a glance. In: Sclaroff, S., Distante, C., Leo, M., Farinella, G.M., Tombari, F. (eds.) ICIAP 2022. LNCS, vol. 13233, pp. 242–253. Springer, Cham (2022). https://doi.org/10.1007/978-3-031-06433-3_21
5. Freire-Obregón, D., Lorenzo-Navarro, J., Santana, O.J., Hernández-Sosa, D., Castrillón-Santana, M.: Towards cumulative race time regression in sports: I3d convnet transfer learning in ultra-distance running events. In: 2022 26th International Conference on Pattern Recognition (ICPR), pp. 805–811 (2022). https://doi.org/10.1109/ICPR56361.2022.9956174

6. Fu, X., Zeng, D., Huang, Y., Liao, Y., Ding, X., Paisley, J.: A fusion-based enhancing method for weakly illuminated images. Sig. Process. **129**, 82–96 (2016). https://doi.org/10.1016/j.sigpro.2016.05.031
7. Guo, X., Li, Y., Ling, H.: LIME: low-light image enhancement via illumination map estimation. IEEE Trans. Image Process. **26**(2), 982–993 (2017). https://doi.org/10.1109/TIP.2016.2639450
8. Hao, S., Han, X., Guo, Y., Wang, M.: Decoupled low-light image enhancement. ACM Trans. Multimedia Comput. Commun. Appl. **18**(4), 1–19 (2022). https://doi.org/10.1145/3498341
9. He, W., Liu, Y., Feng, J., Zhang, W., Gu, G., Chen, Q.: Low-light image enhancement combined with attention map and U-Net network. In: 2020 IEEE 3rd International Conference on Information Systems and Computer Aided Education (ICISCAE), pp. 397–401 (2020). https://doi.org/10.1109/ICISCAE51034.2020.9236828
10. Hernández-Carrascosa, P., Penate-Sanchez, A., Lorenzo-Navarro, J., Freire-Obregón, D., Castrillón-Santana, M.: TGCRBNW: a dataset for runner bib number detection (and recognition) in the wild. In: 2020 25th International Conference on Pattern Recognition (ICPR), pp. 9445–9451 (2021). https://doi.org/10.1109/ICPR48806.2021.9412220
11. Huang, J., et al.: Speed/accuracy trade-offs for modern convolutional object detectors. In: 2017 IEEE Conference on Computer Vision and Pattern Recognition (CVPR), pp. 3296–3297. IEEE Computer Society, Los Alamitos (2017). https://doi.org/10.1109/CVPR.2017.351
12. Jobson, D., Rahman, Z., Woodell, G.: A multiscale retinex for bridging the gap between color images and the human observation of scenes. IEEE Trans. Image Process. **6**(7), 965–976 (1997). https://doi.org/10.1109/83.597272
13. Kim, M.: Improvement of low-light image by convolutional neural network. In: 2019 IEEE 62nd International Midwest Symposium on Circuits and Systems (MWSCAS), pp. 189–192 (2019). https://doi.org/10.1109/MWSCAS.2019.8885098
14. Land, E.H.: The retinex theory of color vision. Sci. Am. **237**(6), 108–129 (1977). http://www.jstor.org/stable/24953876
15. Leng, H., Fang, B., Zhou, M., Wu, B., Mao, Q.: Low-light image enhancement with contrast increase and illumination smooth. Int. J. Pattern Recogn. Artif. Intell. **37**(03), 2354003 (2023). https://doi.org/10.1142/S0218001423540034
16. Li, C., et al.: Low-light image and video enhancement using deep learning: a survey. IEEE Trans. Pattern Anal. Mach. Intell. **44**(12), 9396–9416 (2022). https://doi.org/10.1109/TPAMI.2021.3126387
17. Li, M., Liu, J., Yang, W., Sun, X., Guo, Z.: Structure-revealing low-light image enhancement via robust retinex model. IEEE Trans. Image Process. **27**(6), 2828–2841 (2018). https://doi.org/10.1109/TIP.2018.2810539
18. Liu, J., Xu, D., Yang, W., Fan, M., Huang, H.: Benchmarking low-light image enhancement and beyond. Int. J. Comput. Vis. **129**(4), 1153–1184 (2021). https://doi.org/10.1007/s11263-020-01418-8
19. Liu, S., Long, W., Li, Y., Cheng, H.: Low-light image enhancement based on membership function and gamma correction. Multimedia Tools Appl. **81**, 22087–22109 (2022). https://doi.org/10.1007/s11042-021-11505-8
20. Luo, H., Jiang, W., Zhang, X., Fan, X., Qian, J., Zhang, C.: AlignedReID++: dynamically matching local information for person re-identification. Pattern Recogn. **94**, 53–61 (2019). https://doi.org/10.1016/j.patcog.2019.05.028
21. Lv, F., Lu, F., Wu, J., Lim, C.: MBLLEN: low-light image/video enhancement using CNNs. In: British Machine Vision Conference (BMVC) (2018)
22. Ma, F., Zhu, X., Zhang, X., Yang, L., Zuo, M., Jing, X.Y.: Low illumination person re-identification. Multimedia Tools Appl. **78**, 337–362 (2019). https://doi.org/10.1007/s11042-018-6239-3

23. Ning, X., Gong, K., Li, W., Zhang, L., Bai, X., Tian, S.: Feature refinement and filter network for person re-identification. IEEE Trans. Circ. Syst. Video Technol. **31**(9), 3391–3402 (2021). https://doi.org/10.1109/TCSVT.2020.3043026
24. Park, S., Yu, S., Kim, M., Park, K., Paik, J.: Dual autoencoder network for retinex-based low-light image enhancement. IEEE Access **6**, 22084–22093 (2018). https://doi.org/10.1109/ACCESS.2018.2812809
25. Penate-Sanchez, A., Freire-Obregón, D., Lorenzo-Melián, A., Lorenzo-Navarro, J., Castrillón-Santana, M.: TGC20ReId: a dataset for sport event re-identification in the wild. Pattern Recogn. Lett. **138**, 355–361 (2020). https://doi.org/10.1016/j.patrec.2020.08.003
26. Rahman, Z., Pu, Y.F., Aamir, M., Wali, S.: Structure revealing of low-light images using wavelet transform based on fractional-order denoising and multiscale decomposition. Vis. Comput. **37**(5), 865–880 (2021). https://doi.org/10.1007/s00371-020-01838-0
27. Santana, O.J., Freire-Obregón, D., Hernández-Sosa, D., Lorenzo-Navarro, J., Sánchez-Nielsen, E., Castrillón-Santana, M.: Facial expression analysis in a wild sporting environment. Multimedia Tools Appl. **82**(8), 11395–11415 (2023). https://doi.org/10.1007/s11042-022-13654-w
28. Santana, O.J., Lorenzo-Navarro, J., Freire-Obregón, D., Hernández-Sosa, D., Isern-González, J., Castrillón-Santana, M.: Exploiting temporal coherence to improve person re-identification. In: De Marsico, M., Sanniti di Baja, G., Fred, A. (eds.) ICPRAM 2021, ICPRAM 2022: Pattern Recognition Applications and Methods. LNCS, vol. 13822, pp. 134–151. Springer, Cham (2023). https://doi.org/10.1007/978-3-031-24538-1_7
29. Santana, O.J., Lorenzo-Navarro, J., Freire-Obregón, D., Hernández-Sosa, D., Castrillón-Santana, M.: Evaluating the impact of low-light image enhancement methods on runner re-identification in the wild. In: Proceedings of the 12th International Conference on Pattern Recognition Applications and Methods - ICPRAM, pp. 641–648. INSTICC, SciTePress (2023). https://doi.org/10.5220/0011652000003411
30. Wang, R., Zhang, Q., Fu, C.W., Shen, X., Zheng, W.S., Jia, J.: Underexposed photo enhancement using deep illumination estimation. In: 2019 IEEE/CVF Conference on Computer Vision and Pattern Recognition (CVPR), pp. 6842–6850 (2019). https://doi.org/10.1109/CVPR.2019.00701
31. Wang, S., Zheng, J., Hu, H.M., Li, B.: Naturalness preserved enhancement algorithm for non-uniform illumination images. IEEE Trans. Image Process. **22**(9), 3538–3548 (2013). https://doi.org/10.1109/TIP.2013.2261309
32. Wei, C., Wang, W., Yang, W., Liu, J.: Deep retinex decomposition for low-light enhancement. In: 2018 British Machine Vision Conference (BMVC), pp. 1–12 (2018)
33. Wojke, N., Bewley, A., Paulus, D.: Simple online and realtime tracking with a deep association metric. In: 2017 IEEE International Conference on Image Processing (ICIP), pp. 3645–3649. IEEE Press (2017). https://doi.org/10.1109/ICIP.2017.8296962
34. Ying, Z., Li, G., Ren, Y., Wang, R., Wang, W.: A new low-light image enhancement algorithm using camera response model. In: 2017 IEEE International Conference on Computer Vision Workshops (ICCVW), pp. 3015–3022 (2017). https://doi.org/10.1109/ICCVW.2017.356
35. Zhai, G., Sun, W., Min, X., Zhou, J.: Perceptual quality assessment of low-light image enhancement. ACM Trans. Multimedia Comput. Commun. Appl. **17**(4), 1–24 (2021). https://doi.org/10.1145/3457905
36. Zhang, Q., Nie, Y., Zheng, W.S.: Dual illumination estimation for robust exposure correction. Comput. Graph. Forum **38**(7), 243–252 (2019). https://doi.org/10.1111/cgf.13833
37. Zheng, L., Shen, L., Tian, L., Wang, S., Wang, J., Tian, Q.: Scalable person re-identification: a benchmark. In: 2015 IEEE International Conference on Computer Vision (ICCV), pp. 1116–1124 (2015). https://doi.org/10.1109/ICCV.2015.133

Gender-Aware Speech Emotion Recognition in Multiple Languages

Marco Nicolini and Stavros Ntalampiras(✉)

Department of Computer Science, University of Milan, Milan, Italy
marco.nicolini3@studenti.unimi.it, stavros.ntalampiras@unimi.it

Abstract. This article presents a solution for Speech Emotion Recognition (SER) in multilingual setting using a hierarchical approach. The approach involves two levels, the first level identifies the gender of the speaker, while the second level predicts their emotional state. We evaluate the performance of three classifiers of increasing complexity: k-NN, transfer learning based on YAMNet, and Bidirectional Long Short-Term Memory neural networks. The models were trained, validated, and tested on a dataset that includes the big-six emotions and was collected from well-known SER datasets representing six different languages. Our results indicate that there are differences in classification accuracy when considering all data versus only female or male data, across all classifiers. Interestingly, prior knowledge of the speaker's gender can improve the overall classification performance.

Keywords: Audio pattern recognition · Machine learning · Transfer learning · Convolutional neural network · YAMNet · Multilingual speech emotion recognition

1 Introduction

Speech is a crucial aspect of human-machine communication because it is one of the primary means of expressing emotions. Speech emotion recognition (SER) is a branch of Affective Computing that focuses on automatically identifying a speaker's emotional state through their voice. The ability to sense a user's emotional state is a critical component in developing systems that can interact with humans more naturally [4].

Emotions play an essential role in human behavior and decision-making processes. Emotion recognition has many practical applications, including in the field of healthcare, where it can be used to detect, analyze, and diagnose medical conditions. For instance, SER could be utilized to design medical robots that continuously monitor patients' emotional states and provide more comprehensive and effective healthcare services [23]. Additionally, SER technologies can be employed to develop emotionally-aware human-computer interaction solutions [24].

Most speech emotion recognition (SER) solutions in the literature are designed for a single language, with only a few language-independent methods available, such as those proposed by Saitta [26] and Sharma [20,30]. A significant portion of SER research has focused on identifying speech features that are indicative of different emotions [31]. A

M. De Marsico et al. (Eds.): ICPRAM 2023, LNCS 14547, pp. 111–123, 2024.
https://doi.org/10.1007/978-3-031-54726-3_7

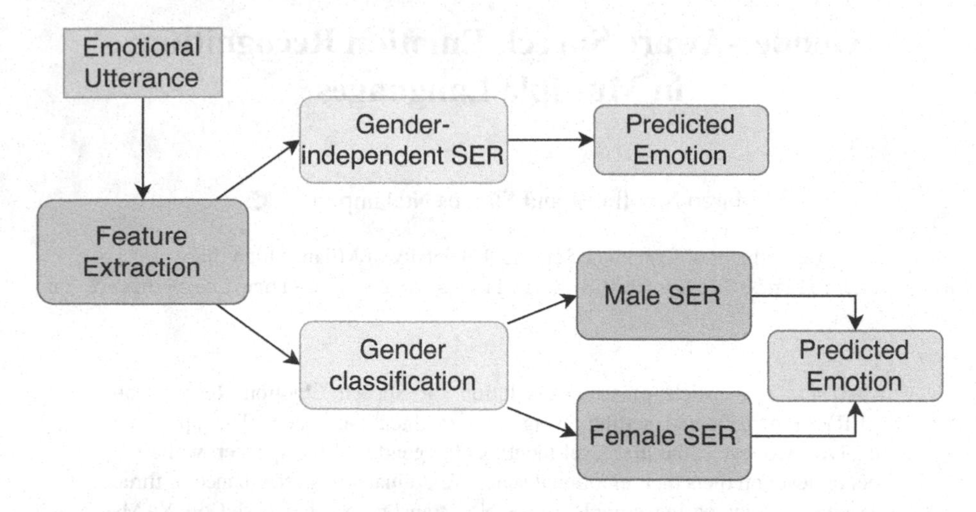

Fig. 1. The block diagram of the classification hierarchy adopted in this work (taken from [17]).

range of both short-term and long-term features have been proposed for this purpose. Emotions are typically classified into six basic categories: angry, disgust, happy, sad, neutral, and fear [14].

Identifying the correct classification boundaries for speech emotions can be challenging due to overlapping features between emotions. Deep learning methods provide a promising solution as they can automatically discover multiple levels of representations in speech signals [28]. Consequently, there is increasing interest in applying deep learning-based methods to learn useful features from emotional speech data. For example, Mirsamadi [15] applies recurrent neural networks to discover emotionally relevant features and classify emotions, while Kun Han [8] uses Deep Neural Networks for SER. The work of Scheidwasser et al. proposes a framework to evaluate the performance and generalization capacity of different approaches for SER, mainly deep learning methods. However, their approach is language-dependent, training different models on one dataset of the six benchmark datasets at a time [29].

The proposed methodology in this work focuses on a language-agnostic approach for SER, aiming to generalize patterns in data to distinguish emotions across all languages in the dataset. The proposed solution includes automatic gender differentiation, which improves the SER performance of the classifiers. This approach is also explored by Vogt et al. [34], who present a framework to improve emotion recognition from monolingual speech using automatic gender detection. Dair [6] also investigates the impact of gender differentiation on three datasets.

In contrast to previous studies in the literature, our proposed system takes into account all six big emotions in a multilingual context that involves six different languages. Based on the findings of our previous work [17], this paper presents and analyzes the results obtained following the proposed SER approach while considering each language independently. As such, we detail the differences in performance observed in each language and provide a thorough comparison between gender-aware and gender-agnostic classification approaches.

The block diagram of the proposed approach is demonstrated in Fig. 1 where the following three classifiers can be observed:

- gender,
- female, and
- male emotional speech.

The gender-independent path is shown as well since it has been employed for comparison purposes.

The system employs three different classifiers for emotion recognition: a k-Nearest Neighbor Classifier, a transfer learning-based classifier using YAMNet, and a Bidirectional Long Short-Term Memory neural network Classifier. While the k-NN classifier is simple, it is still suitable for multi-class problems [9]. The latter two classifiers belong to the deep learning field, with the transfer learning-based classifier relying on a large-scale convolutional network and the BiLSTM classifier being capable of encoding temporal dependencies in the emotional expressions. All models were trained on appropriate features extracted from the time and/or frequency domains.

The remainder of this paper is structured as follows: Sect. 2 describes the development of a multilingual dataset for SER purposes. Section 3 briefly discusses the features and classifiers used in the system. Section 4 presents the experimental setup and results, while Sect. 5 provides conclusions and possible future research directions.

Table 1. Duration (in seconds) of the diverse classes considered in the present work. All values are truncated (taken from [17]).

Data part	Angry	Neutral	Sad	Happy	Fear	Disgust
All Data (50541)	10594	11200	9325	7116	5773	6530
Female (25081)	5023	4706	4948	3807	3018	3576
Male (25460)	5570	6493	4377	3309	2754	2953
English (32154)	5475	5464	5772	5174	4765	5502
German (1261)	335	186	251	180	154	154
Italian (1711)	268	261	313	266	283	317
Urdu (998)	248	250	250	250	0	0
Persian (11932)	3832	5037	2175	766	120	0
Greek (2481)	433	0	562	478	449	556

2 Constructing the Multilingual SER Dataset

In order to create a multilingual dataset for SER purposes, we combined ten different monolingual datasets commonly used in the SER literature.

- SAVEE [33],
- CREMA-D [2],
- RAVDESS [13],
- TESS [25],
- EMOVO [5],
- EmoDB [1],
- ShEMO [16],
- URDU [11],
- JLcorpus [10], and
- AESDD [35,36].

These datasets cover the 'big six' emotions that are typically considered in SER research [14]. The following six languages are considered in the final dataset:

- English (New Zealand, British, American, and from different ethnic backgrounds),
- German,
- Italian,
- Urdu,
- Persian, and
- Greek.

The duration in seconds for each class included in the dataset is presented in Table 1. It is worth noting that all datasets comprise acted speech. There is only a slight imbalance in terms of gender, with male emotional speech lasting for 25460 s and female emotional speech lasting for 25081 s. Moreover, the English language makes up the largest portion of the dataset, followed by Persian, while the speech data of the remaining languages (German, Italian, Greek, and Urdu) range from 998 to 2481 s.

Aiming at a uniform representation, all data has been resampled to 16 kHz and monophonic wave format. In constructing a SER system, several challenges arise due to differences in languages and cultures, for instance:

- different languages presenting important cultural gaps,
- imbalances at the genre, language, and emotional state levels,
- diverse recording conditions, and
- different recording equipment.

To address the first two challenges, the data was appropriately divided during the training, validation, and testing phases to ensure that the resulting models are not biased toward any sub-population within the corpus. Regarding the last two challenges, the proposed approach aims to create a standardized representation of the audio signal to minimize the effect of recording conditions and equipment.

To the best of our knowledge, this is the first time that the entire range of big-six emotional states expressed in six different languages has been considered in the SER literature.

3 The Considered Classification Models

In this section, we provide a brief overview of the classification models used in this study and their respective feature sets that capture the characteristics of the available audio data. Notably, each classification model has been trained and evaluated separately on three different settings:

- the entire dataset,
- female data, and
- male data.

The same data split into train, validation, and test sets has been used for all classifiers to ensure a fair comparison among them. Additionally, the k-NN and YAMNet-based models have been specifically trained to differentiate between male and female speech.

3.1 k-NN

The k-NN classifier was implemented in its standard form with the Euclidean distance serving as the similarity metric. Although it is a simple classifier, k-NN has shown to perform well in SER [32]. Therefore, we evaluated its performance on the current challenging multilingual task.

Feature Extraction. The short-term features [21] feeding the k-NN model are the following:

- zero crossing rate,
- energy,
- energy's entropy,
- spectral centroid and spread,
- spectral entropy,
- spectral flux,
- spectral rolloff,
- MFCCs,
- harmonic ratio,
- fundamental frequency, and
- chroma vectors.

The mid-term feature extraction process was chosen, which involves calculating mean and standard deviation statistics on short-term features over mid-term segments. Further details on the feature extraction method used can be found in [7].

Parameterization. The appropriate window and hop sizes for short- and mid-term feature extraction were determined through a series of preliminary experiments on the different datasets. The configuration that yielded the highest accuracy was found to be: 0.2 and 0.1 s for short-term window and hop size, respectively, and 3 and 1.5 s for mid-term window and hop size, respectively. Both feature extraction levels included a 50% overlap between adjacent windows.

Furthermore, the optimal value for parameter k was determined using test results based on the ten-fold cross-validation scheme. Depending on the population of the data being considered, the optimal values ranged from 5 to 21 (refer to Sect. 4 for further details).

Table 2. Confusion matrix (in %) as regards to gender classification obtained using the YAMNet-based and k-NN approaches. The presentation format is the following: YAMNet/k-NN. The highest accuracy is emboldened (taken from [17]).

Presented	Predicted	
	Female	*Male*
Female	**94.3**/91.1	5.7/8.9
Male	4.8/3.6	95.2/**96.4**

3.2 Transfer Learning Based on YAMNet

YAMNet is a neural network model developed by Google that has been trained on 512 classes of generalized audio events from the AudioSet ontology[1]. Because of this, the learned representation can be beneficial for various audio classification tasks, including SER. For this purpose, we focused on the Embeddings layer of the model and utilized it as a feature set that serves as input to a dense layer with a number of neurons equal to the classes that need to be classified (2 genres or 6 emotions). The final prediction is made using a softmax layer, and a dictionary of weights can also be applied to address imbalanced data classes.

3.3 Bidirectional LSTM

The inclusion of a Long Short-Term Memory (LSTM) network in the experimental set-up is motivated by its effectiveness in capturing long-term temporal dependencies, which is particularly relevant in the case of audio signals that evolve over time. Specifically, we opted for a bidirectional LSTM (BiLSTM) layer, which learns bidirectional long-term dependencies in sequential data. BiLSTMs are an extension of traditional LSTMs that have shown to improve model performance on sequence classification problems [27]. In a BiLSTM, two LSTMs are trained on the input sequence, with the first LSTM processing the sequence as-is and the second LSTM processing a reversed copy of the input sequence.

Feature Extraction. In this case, the considered feature sets, able to preserve the temporal evolution of the available emotional manifestations, were the following:

- Gammatone cepstral coefficients (GTCC),
- Delta Gammatone cepstral coefficients (delta GTCC),
- delta-delta MFCC,
- Mel spectrurm, and
- spectral crest.

The window length has been chosen after extensive experimentations performed on the different data subpopulations, e.g. genre, emotions, etc., while there is no overlapping between subsequent windows.

[1] https://research.google.com/audioset/index.html.

Table 3. Average classification accuracy and balanced accuracy results (in %) using 10 fold evaluation. The presentation format is the following: accuracy/balanced accuracy. The highest accuracy and balanced accuracy for each data subpopulation are emboldened (taken from [17]).

Data subpopulation	k-NN	YAMNet	BiLSTM
all data	59/59	**74.9**/46.9	62/**59.1**
female data	65.2/65.1	**79.9**/52	68.6/**67**
male data	51.6/**51.6**	**71.7**/47.1	56.7/51.3

Table 4. Confusion matrix (in %) as regards to SER classification obtained using the BiLSTM, k-NN and YAMNet models approaches with all data. The presentation format is the following: BiLSTM/k-NN/YAMNet. The highest accuracy is emboldened (taken from [17]).

Pres.	Pred.					
	angry	*disgust*	*fear*	*happy*	*neutral*	*sad*
angry	**83.1**/68.7/60.9	3.8/8.9/12.1	0.9/3.2/23.8	5.8/13.4/3.2	5.5/4.8/-	0.9/0.9/-
disgust	11.6/5/1.4	45/64.3/**76.6**	4.3/7.4/19.3	4.6/8/2.8	16.8/7.3/-	17.7/8/-
fear	10.5/6/0.6	6.8/17.6/5.1	41.2/47.4/**94.3**	8.6/10.7/-	7.6/5.2/-	25.3/13.1/-
happy	22/13.3/14.8	4.9/14.9/34.4	8.8/8.8/26.2	43.2/**50.2**/24.6	16.1/9.2/-	5/3.5/-
neutral	2.8/2.7/-	5.3/12.3/40	1/4.6/31.4	3.2/4.6/17.1	**75.6**/67.3/8.6	12.1/8.6/2.9
sad	1.5/3.5/-	3/11.5/72	4/11.5/12	1.8/3.7/16	24.1/13.6/-	**65.7**/56.2/-

4 Experimental Protocol and Results

We conducted our experiments following the 10-fold cross-validation protocol, ensuring that each classifier was trained, validated, and tested on identical folds. The average accuracy achieved by each classifier is presented in Table 3, while confusion matrices for gender-independent and dependent SER are shown in Tables 4, 5, 6 and 7. Based on these results, we make the following observations.

Firstly, both YAMNet and k-NN perform well in gender discrimination, with YAMNet offering the highest accuracy, as shown in the respective confusion matrix (Table 2). Notably, these results are comparable to the state of the art in gender discrimination [3].

Secondly, in terms of SER, YAMNet achieves the highest unbalanced accuracy, but the confusion matrices in Tables 4, 5, 6 and 7 reveal that its performance varies across emotional states, performing well for some (e.g., *fear*) and poorly for others (e.g., *happy*).

Thirdly, the BiLSTM models outperform the other classifiers, with an average accuracy of 62% across all data and per-class accuracy measures not falling below 56.1% (Table 4). This may be due to their ability to capture temporal dependencies, which are important for speech processing in general [12]. Additionally, BiLSTM models trained on male or female data provide satisfactory performance (Tables 6, 7), with a balanced accuracy of 62.7%.

Fourthly, the k-NN models perform comparably to the BiLSTM models, with the associated confusion matrices showing its capacity to distinguish between various emotional classes. The optimal k values obtained for the all-data, female-data, male-data, and gender-data models are $k = 11$, $k = 21$, $k = 13$, and $k = 5$, respectively. These

Table 5. Table summarizing the accuracies (in %) reached by the considered classifiers when trained on every language independently. Presentation format: k-nn/bilstm/Yamnet

Language	Data		
	All data	Male data	Female data
English	58.8/60.6/50.4	47.5/49.7/43.7	67.9/69.1/56.4
German	70.6/70.3/69.1	78.1/62.6/69.4	75.2/70.3/74.3
Italian	64.8/56.5/53.7	66.3/60.7/56.8	71.7/62.7/64.5
Persian	52.2/76.8/71.2	51.6/78.9/74.9	52.5/72.4/69.8
Greek	57.4/68.3/52.2	68.3/65.7/57.9	53.4/62.3/53.3
Urdu	82.3/82.2/85.8	81.8/83.1/86.7	95.2/85.7/92

results are consistent with findings that distributed modeling techniques may be effective in multilingual settings [19].

Lastly, our gender-dependent classification results follow a pattern observed in prior literature (e.g., Vogt [34]), with performance improving when female emotional speech is considered but not for male. As gender discrimination achieved almost perfect accuracy (over 94%), we suggest using a hierarchical classifier that combines gender and emotion recognition, which can improve the overall recognition rate of gender-independent SER, especially when considering female utterances.

In order to enable reliable comparison with other solutions and full reproducibility, the implementation of the experiments presented in this paper is publicly available at https://github.com/NicoRota-0/SER.

4.1 Performance Analysis of Each Classifier Across Different Languages

To further evaluate the performance of the proposed SER algorithm, we conducted experiments on the six languages, i.e. English, German, Italian, Persian, Greek and Urdu, independently. We trained the k-NN, YAMNet, and BILSTMs models on all data, as well as male and female data for each language. The summary of all these model's accuracy results are presented in Table 5.

There, we can observe that the accuracy of the classifiers varies significantly across languages and gender. For example, the k-NN model achieved the highest accuracy on Urdu, while the BiLSTMs model performed the best on German. Interestingly, we found that the gender-based emotion classifier outperformed the general emotion classifier for female data in most cases. For example, on the Urdu dataset, the accuracy of the female data was much higher than that of the male data for all three models.

These results suggest that the proposed SER algorithm is capable of achieving high accuracy for emotion classification across different languages and gender, although the performance may vary depending on the specific dataset and classifier used. In fact it is important to note that the language specific datasets are very unbalanced, and this adds complexity for the classification problem; also, reducing data for training the models leads to less information for the models to distinguish between the various emotions.

Based on the results, it appears that the accuracy of the models varies depending on the language of the audio data. Specifically, the results show that the BILSTMs model outperforms the other models on English and Persian (see confusion matrices in Tables

Table 6. Confusion matrix (in %) as regards to SER classification obtained using the BiLSTM, k-NN and YAMNet models approaches with **female** data. The presentation format is the following: BiLSTM/k-NN/YAMNet. The highest accuracy is emboldened (taken from [17]).

Pres.	Pred.					
	angry	*disgust*	*fear*	*happy*	*neutral*	*sad*
angry	**85.4**/75.5/63.2	2.4/6.6/16.1	0.9/2.7/19.5	6.6/9.9/1.1	3.6/4/-	1.1/1.3/-
disgust	9.1/7.8/0.8	56.7/67.4/**88.4**	2/4.1/10.1	4.7/5.3/0.8	13.5/7/-	13.9/7/-
fear	7.8/6.7/0.4	5.3/9.4/3.1	51.6/58.7/**96.5**	8.2/7.6/-	5.3/11.3/-	21.9/11.3/-
happy	16.2/15.5/14.8	4.5/9.7/59.3	6.5/8.7/25.9	**57.8**/54.8/-	10.1/8/-	4.8/3.3/-
neutral	2.7/3.4/-	8/12.1/40	0.5/2.4/50	2.9/2.5/-	**74.8**/71.7/10	11.1/7.8/-
sad	1.8/4.5/-	3.1/7.4/71.4	3.6/8.6/28.6	1.3/2.7/-	14.6/13.9/-	**75.5**/63/-

Table 7. Confusion matrix (in %) as regards to SER classification obtained using the BiLSTM, k-NN and YAMNet models approaches with **male** data. The presentation format is the following: BiLSTM/k-NN/YAMNet. The highest accuracy is emboldened (taken from [17]).

Pres.	Pred.					
	angry	*disgust*	*fear*	*happy*	*neutral*	*sad*
angry	**80.6**/64/58.4	3.8/10.5/11.2	0.7/4.6/25.5	6.9/13.6/5	7.2/6.2/-	0.9/1.1/-
disgust	13.8/4.9/1.9	33/55.7/**66.5**	6.5/10.5/27.3	7.2/10.9/4.3	17.4/6.9/-	22/11.1/-
fear	14.7/5.7/0.3	6.9/23.8/4.4	30.4/34.6/**95.3**	11.1/11.5/-	9.1/6.5/-	27.7/18/-
happy	26.1/13/11.8	6.1/18.5/23.5	11.5/10.8/23.5	31.4/**41.2**/38.2	19.9/12.3/-	5/4.3/2.9
neutral	2.7/2.4/-	5.1/11.6/56	1.9/5.2/28	3.5/5.5/16	**75.8**/65.5/-	11/9.6/-
sad	2.2/1.8/-	3.6/14.6/61.1	5.3/13/-	2.5/4.6/33.3	29.6/17.1/-	**56.7**/48.8/5.6

Table 8. Confusion matrix (in %) as regards to SER classification obtained using the BiLSTMs model approach for **English**. The presentation format is the following: all/male/female.

Pres.	Pred.					
	angry	*disgust*	*fear*	*happy*	*neutral*	*sad*
angry	**80/74.3/86**	5/7.9/3	2/2/1	9/10.9/7	3/4/3	1/1/0
disgust	11/13.9/9	**48/32.7/57**	5/7.9/2	5/5/4	14/19.8/14	17/20.8/14
fear	9/13.9/5.9	7/9.9/4	**44**/23.8/**57.4**	9/13.9/7.9	8/10.9/5.9	23/**27.7**/18.8
happy	18.8/24/16	5/7/5	9.9/14/8	**50.5/36/60**	11.9/15/9	4/4/2
neutral	1/1/1	8.1/10/7	2/3/1	5.1/7/3	**68.7/66/77**	15.2/13/11
sad	1/1/1	5/4/4	5.9/8/5	2/2/1	17.8/27/14.9	**68.3/58/74.3**

Table 9. Confusion matrix (in %) as regards to SER classification obtained using the k-NN model approach for **German**. The presentation format is the following: all/male/female.

Pres.	Pred.					
	angry	*neutral*	*sad*	*happy*	*fear*	*disgust*
Angry	**80.5/78/71.3**	0/0.8/0	0/0/0	13/12.2/19.9	2.7/5.7/2.9	3.8/3.3/5.9
Neutral	0/0/0	**78.8/91.2/78.9**	1.3/3.8/10.5	1.9/0/0	7.1/2.5/0	10.9/2.5/10.5
Sad	0/0/0	12.6/14.3/3.3	**86.5/83.7/93.4**	0/0/0	0/0/0	0.9/2/3.3
Happy	26.9/15.9/31.7	3.4/11.4/0	0/0/0	**53.8/70.5/51.2**	2.1/2.3/7.3	13.8/0/9.8
Fear	13.5/7.9/11.3	6.4/23.7/1.9	1.3/1.3/1.9	5.8/3.9/3.8	**51.3/59.2/71.7**	21.8/3.9/9.4
Disgust	6.2/0/6.9	7.4/10.7/2.8	6.2/0/2.8	2.5/0/1.4	4.9/3.6/1.4	**72.8/85.7/84.7**

Table 10. Confusion matrix (in %) as regards to SER classification obtained using the k-NN model approach for **Italian**. The presentation format is the following: all/male/female.

Pres.	Pred.					
	angry	*neutral*	*sad*	*happy*	*fear*	*disgust*
angry	**72.8/67.9/76.2**	0.6/2.6/0	0/2.6/0	16/15.4/18.8	3.7/2.6/5	6.8/9/0
neutral	5.7/2.2/0	**70.1/82/94.8**	1.9/0/0	7/3.4/2.6	3.8/6.7/0	11.5/5.6/2.6
sad	2.3/2.5/0	2.9/6.2/0	**73.1/81.5/90.6**	0.6/0/0	12.6/4.9/9.4	8.6/4.9/0
happy	19/18.8/13.4	7.1/8.2/8.2	0.6/0/0	**48.2/55.3/46.4**	9.5/10.6/10.3	15.5/7.1/21.6
fear	4.2/6.3/2.1	2.4/12.6/0	7.1/1.1/9.5	12.5/12.6/1.1	**63.1/52.6/76.8**	10.7/14.7/10.5
disgust	6.7/6.9/3	5/6.9/9.1	11.1/11.1/13.6	8.3/1.4/16.7	7.2/15.3/12.1	**61.7/58.3/45.5**

Table 11. Confusion matrix (in %) as regards to SER classification obtained using the BiLSTMs model approach for **Persian**. The presentation format is the following: all/male/female.

Pres.	Pred.				
	angry	*fear*	*happy*	*neutral*	*sad*
angry	**88/86/90**	0/0/0	0/0/0	11/14/6	1/0/4
fear	29/25/32	0/0/0	0/6/0	34/44/9	**37/25/59**
happy	28.3/27/35	0/0/0	5.1/7/5	**46.5/64/26**	20.2/2/34
neutral	5.1/4/11.9	0/0/0	0/0/1	**92.9/96/76.2**	2/0/10.9
sad	6/4/8	0/1/0	1/1/1	39/64.4/19	**54/29.7/72**

Table 12. Confusion matrix (in %) as regards to SER classification obtained using the BiLSTMs model approach for **Greek**. The presentation format is the following: all/male/female.

Pres.	Pred.				
	angry	*disgust*	*fear*	*happy*	*sad*
angry	**79/57/77**	10/22/3	2/0/0	7/18/10	2/3/10
disgust	2/3/3	**84/85/69**	7/5/5	2/2/0	5/5/23
fear	5/6/5	26.7/27/28	**41.6/44/30**	9.9/10/16	16.8/13/21
happy	22/17/18	22/24/28	11/8/7	**40/46/40**	5/5/7
sad	0/0/2	5/0/5	1/3/2	0/0/0	**94/97/91.1**

Table 13. Confusion matrix (in %) as regards to SER classification obtained using the YAMNet model approach for **Urdu**. The presentation format is the following: all/male/female.

Pres.	Pred.			
	angry	*happy*	*neutral*	*sad*
angry	**93.2/93.9/-**	3.4/1.4/-	2.7/2.7/-	0.7/2/-
happy	3.3/4.9/-	**82.7/69.1/100**	10.7/18.5/0	3.3/7.4/0
neutral	3.3/1.4/-	10/7.6/50	**82.7/81.9/50**	4/9/0
sad	0.7/1.4/-	8.7/2.8/33.3	6/1.4/0	**84.7/94.3/66.7**

8, 11), while the k-NN model performs best on German and Italian (see confusion matrices in Tables 9, 10) and appears to perform relatively consistently across languages. For Greek and Urdu, as it is possible to see in Table 5 that the best accuracies are reached by different models on all data, male and female data. For Greek: BiLSTM model is the best-performing one for all data and female data (confusion matrix in Table 12), on the other hand for male the best-performing is k-nn. For Urdu language: the best-performing model for all data and male data is YAMNet (confusion matrix in Table 13), while for female it is k-nn.

When comparing the results for the multilingual dataset versus the language-specific datasets, we see that the YAMNet model performs significantly better on the multilingual dataset than on any individual language, especially in terms of accuracy. This suggests that YAMNet is better at recognizing emotions in a multilingual context than in any individual language. However, for the other models, the results show that they tend to perform better on individual languages than on the multilingual dataset. This could be due to variations in the way emotional expressions are conveyed across different languages, which may require more specialized models to recognize. Overall, the results suggest that while some models may perform well in a multilingual context, for other models it may be more effective to train separate models for each language in order to achieve the best possible accuracy.

Further analysis and experimentation could be conducted to better understand the factors that influence the performance of SER algorithms and how they can be optimized for specific applications.

5 Conclusion and Future Developments

This study explores multilingual audio gender-based emotion classification, presenting a novel SER algorithm that achieves state-of-the-art results while considering the full range of the big six emotional states expressed in six languages. Interestingly, the findings indicate that a gender-based emotion classifier can outperform a general emotion classifier.

While this study demonstrates promising results, there are still avenues for future research. One potential direction is to assess the performance of these modeling architectures on each language separately, as this could yield insights into any potential language-specific trends or challenges. Additionally, these models could be incorporated into more complex systems that use biosensors to measure physiological parameters (such as heart rate) to more accurately detect and classify emotions. This is particularly relevant given the accelerating spread of IoT devices, as noted in Pal's work [22].

Another potential area for future research is investigating the effectiveness of the one-vs-all emotion classification scheme using the models developed in this study. This approach has been explored in the work of Saitta et al. [26], and may yield useful insights into the performance of these models across different emotional states. Alternatively, researchers could experiment with incorporating a language classifier before emotion detection (with or without gender detection) to explore whether this approach yields improved results. Finally, another line of work will include the exploration of transfer learning technologies [18] where a language-specific SER models may be adapted from large language-agnostic ones.

Acknowledgements. This work was carried out within the project entitled "Advanced methods for sound and music computing" funded by the University of Milan.

References

1. Burkhardt, F., Paeschke, A., Rolfes, M., Sendlmeier, W.F., Weiss, B., et al.: A database of German emotional speech. In: Interspeech, vol. 5, pp. 1517–1520 (2005)
2. Cao, H., Cooper, D.G., Keutmann, M.K., Gur, R.C., Nenkova, A., Verma, R.: Crema-d: crowd-sourced emotional multimodal actors dataset. IEEE Trans. Affect. Comput. **5**(4), 377–390 (2014)
3. Chachadi, K., Nirmala, S.R.: Voice-based gender recognition using neural network. In: Joshi, A., Mahmud, M., Ragel, R.G., Thakur, N.V. (eds.) Information and Communication Technology for Competitive Strategies (ICTCS 2020). LNNS, vol. 191, pp. 741–749. Springer, Singapore (2021). https://doi.org/10.1007/978-981-16-0739-4_70
4. Chen, L., Wang, K., Li, M., Wu, M., Pedrycz, W., Hirota, K.: K-means clustering-based kernel canonical correlation analysis for multimodal emotion recognition in human-robot interaction. IEEE Trans. Industr. Electron. **70**(1), 1016–1024 (2023). https://doi.org/10.1109/TIE.2022.3150097
5. Costantini, G., Iaderola, I., Paoloni, A., Todisco, M.: EMOVO corpus: an Italian emotional speech database. In: International Conference on Language Resources and Evaluation (LREC 2014), pp. 3501–3504. European Language Resources Association (ELRA) (2014)
6. Dair, Z., Donovan, R., O'Reilly, R.: Linguistic and gender variation in speech emotion recognition using spectral features. IEEE Signal Process. Lett. **29**, 250–254 (2022)
7. Giannakopoulos, T., Pikrakis, A.: Introduction to Audio Analysis: A MATLAB Approach, 1st edn. Academic Press Inc, USA (2014)
8. Han, K., Yu, D., Tashev, I.: Speech emotion recognition using deep neural network and extreme learning machine. In: Interspeech 2014 (2014)
9. Hota, S., Pathak, S.: KNN classifier based approach for multi-class sentiment analysis of twitter data. Int. J. Eng. Technol. **7**(3), 1372 (2018). https://doi.org/10.14419/ijet.v7i3.12656
10. James, J., Tian, L., Watson, C.I.: An open source emotional speech corpus for human robot interaction applications. In: INTERSPEECH, pp. 2768–2772 (2018)
11. Latif, S., Qayyum, A., Usman, M., Qadir, J.: Cross lingual speech emotion recognition: Urdu vs. western languages. In: 2018 International Conference on Frontiers of Information Technology (FIT), pp. 88–93. IEEE (2018)
12. Latif, S., Rana, R., Khalifa, S., Jurdak, R., Schuller, B.W.: Self supervised adversarial domain adaptation for cross-corpus and cross-language speech emotion recognition. IEEE Trans. Affect. Comput. 1–1 (2022). https://doi.org/10.1109/TAFFC.2022.3167013
13. Livingstone, S.R., Russo, F.A.: The ryerson audio-visual database of emotional speech and song (ravdess): a dynamic, multimodal set of facial and vocal expressions in north American English. PLoS ONE **13**(5), e0196391 (2018)
14. Miller, H.L., Jr.: The Sage Encyclopedia of Theory in Psychology. SAGE Publications, Thousand Oaks (2016)
15. Mirsamadi, S., Barsoum, E., Zhang, C.: Automatic speech emotion recognition using recurrent neural networks with local attention. In: 2017 IEEE International Conference on Acoustics, Speech and Signal Processing (ICASSP), pp. 2227–2231. IEEE (2017)
16. Nezami, O.M., Lou, P.J., Karami, M.: ShEMO: a large-scale validated database for Persian speech emotion detection. Lang. Resour. Eval. **53**(1), 1–16 (2019)
17. Nicolini, M., Ntalampiras, S.: A hierarchical approach for multilingual speech emotion recognition. In: Proceedings of the 12th International Conference on Pattern Recognition Applications and Methods. SCITEPRESS - Science and Technology Publications (2023). https://doi.org/10.5220/0011714800003411
18. Ntalampiras, S.: Bird species identification via transfer learning from music genres. Eco. Inform. **44**, 76–81 (2018). https://doi.org/10.1016/j.ecoinf.2018.01.006

19. Ntalampiras, S.: Toward language-agnostic speech emotion recognition. J. Audio Eng. Soc. **68**(1/2), 7–13 (2020). https://doi.org/10.17743/jaes.2019.0045
20. Ntalampiras, S.: Speech emotion recognition via learning analogies. Pattern Recogn. Lett. **144**, 21–26 (2021)
21. Ntalampiras, S.: Model ensemble for predicting heart and respiration rate from speech. IEEE Internet Comput. 1–7 (2023). https://doi.org/10.1109/MIC.2023.3257862
22. Pal, S., Mukhopadhyay, S., Suryadevara, N.: Development and progress in sensors and technologies for human emotion recognition. Sensors **21**(16), 5554 (2021). https://doi.org/10.3390/s21165554
23. Park, J.S., Kim, J.H., Oh, Y.H.: Feature vector classification based speech emotion recognition for service robots. IEEE Trans. Consum. Electron. **55**(3), 1590–1596 (2009)
24. Pavlovic, V., Sharma, R., Huang, T.: Visual interpretation of hand gestures for human-computer interaction: a review. IEEE Trans. Pattern Anal. Mach. Intell. **19**(7), 677–695 (1997). https://doi.org/10.1109/34.598226
25. Pichora-Fuller, M.K., Dupuis, K.: Toronto emotional speech set (TESS). Scholars Portal Dataverse **1**, 2020 (2020)
26. Saitta, A., Ntalampiras, S.: Language-agnostic speech anger identification. In: 2021 44th International Conference on Telecommunications and Signal Processing (TSP), pp. 249–253. IEEE (2021)
27. Sajjad, M., Kwon, S.: Clustering-based speech emotion recognition by incorporating learned features and deep BiLSTM. IEEE Access **8**, 79861–79875 (2020). https://doi.org/10.1109/ACCESS.2020.2990538
28. Sang, D.V., Cuong, L.T.B., Ha, P.T.: Discriminative deep feature learning for facial emotion recognition. In: 2018 1st International Conference on Multimedia Analysis and Pattern Recognition (MAPR), pp. 1–6 (2018). https://doi.org/10.1109/MAPR.2018.8337514
29. Scheidwasser-Clow, N., Kegler, M., Beckmann, P., Cernak, M.: SERAB: a multi-lingual benchmark for speech emotion recognition. In: ICASSP 2022–2022 IEEE International Conference on Acoustics, Speech and Signal Processing (ICASSP), pp. 7697–7701. IEEE (2022)
30. Sharma, M.: Multi-lingual multi-task speech emotion recognition using wav2vec 2.0. In: Proceedings of the 2022 IEEE International Conference on Acoustics, Speech and Signal Processing (ICASSP), pp. 6907–6911. IEEE (2022)
31. Tahon, M., Devillers, L.: Towards a small set of robust acoustic features for emotion recognition: challenges. IEEE/ACM Trans. Audio, Speech, Lang. Process. **24**(1), 16–28 (2015)
32. Venkata Subbarao, M., Terlapu, S.K., Geethika, N., Harika, K.D.: Speech emotion recognition using k-nearest neighbor classifiers. In: Shetty D., P., Shetty, S. (eds.) Recent Advances in Artificial Intelligence and Data Engineering. AISC, vol. 1386, pp. 123–131. Springer, Singapore (2022). https://doi.org/10.1007/978-981-16-3342-3_10
33. Vlasenko, B., Schuller, B., Wendemuth, A., Rigoll, G.: Combining frame and turn-level information for robust recognition of emotions within speech. In: Proceedings of Interspeech, pp. 2249–2252 (2007)
34. Vogt, T., André, E.: Improving automatic emotion recognition from speech via gender differentiation. In: Proceedings of the 5th Language Resources and Evaluation Conference (LREC), pp. 1123–1126 (2006)
35. Vryzas, N., Kotsakis, R., Liatsou, A., Dimoulas, C.A., Kalliris, G.: Speech emotion recognition for performance interaction. J. Audio Eng. Soc. **66**(6), 457–467 (2018)
36. Vryzas, N., Matsiola, M., Kotsakis, R., Dimoulas, C., Kalliris, G.: Subjective evaluation of a speech emotion recognition interaction framework. In: Proceedings of the Audio Mostly 2018 on Sound in Immersion and Emotion, pp. 1–7. Association for Computing Machinery (2018)

Pattern Recognition Techniques in Image-Based Material Classification of Ancient Manuscripts

Maruf A. Dhali[1(✉)], Thomas Reynolds[2], Aylar Ziad Alizadeh[1], Stephan H. Nijdam[1], and Lambert Schomaker[1]

[1] Department of Artificial Intelligence, University of Groningen, Groningen, The Netherlands
{m.a.dhali,l.r.b.schomaker}@rug.nl,
{f.ziad.alizadeh,s.h.nijdam}@student.rug.nl

[2] Department of Computer Science, Royal Holloway, University of London, London, UK
thomas.reynolds.2021@live.rhul.ac.uk

Abstract. Classifying ancient manuscripts based on their writing surfaces often becomes essential for palaeographic research, including writer identification, manuscript localization, date estimation, and, occasionally, forgery detection. Researchers continually perform corroborative tests to classify manuscripts based on physical materials. However, these tests, often performed on-site, require actual access to the manuscript objects. These procedures involve specific expertise in manuscript handling, a considerable amount of time, and cost. Additionally, any physical inspection can accidentally damage ancient manuscripts that already suffer degradation due to aging, natural corrosion, and damaging chemical treatments. Developing a technique to classify such documents using noninvasive techniques with only digital images can be extremely valuable and efficient. This study uses images from a famous historical collection, the Dead Sea Scrolls, to propose a novel method to classify the materials of the manuscripts. The proposed classifier uses the two-dimensional Fourier transform to identify patterns within the manuscript surfaces. Combining a binary classification system employing the transform with a majority voting process is adequate for this classification task. This initial study shows a classification percentage of up to 97% for a confined amount of manuscripts produced from either parchment or papyrus material. In the extended work, this study proposes a hierarchical k-means clustering method to group image fragments that are highly likely to originate from a similar source using color and texture features calculated on the image patches, achieving 77% and 68% for color and texture clustering with 100% accuracy on primary material classification. Finally, this study explores a convolutional neural network model in a self-supervised Siamese setup with a large number of images that obtains an accuracy of 85% on the pretext task and an accuracy of 66% on the goal task to classify the materials of the Dead Sea Scrolls images.

Keywords: Document analysis · Image-based material analysis · Historical manuscript · Feature extraction · Fourier transform · Classification · Clustering · Convolutional neural network

1 Introduction

In palaeographic research, the classification of ancient manuscripts according to their writing surfaces is significant because it may be utilized to identify the writers, locate

M. De Marsico et al. (Eds.): ICPRAM 2023, LNCS 14547, pp. 124–150, 2024.
https://doi.org/10.1007/978-3-031-54726-3_8

the manuscripts, estimate the dates, and look for possible forgeries, among other things. Researchers typically perform tests that evaluate the surface materials used to create these manuscripts to classify them appropriately. However, carrying out these tests frequently necessitates having direct access to the document objects, which can be a practical hassle. Such access requires specialist knowledge in managing manuscripts, takes a lot of time, and costs a lot of money. Furthermore, due to the fragile nature of these old manuscripts, physical inspection has inherent risks. The manuscripts are vulnerable to deterioration brought on by things like natural corrosion, accumulated chemical treatments [17], and everyday wear and tear [13]. Therefore, there is always a risk of accidental damage during any physical inspection performed for research purposes, further jeopardizing the integrity and preservation of the manuscripts.

Additionally, gaining first-hand access to such manuscripts is often restricted or impractical. Non-invasive methods for classifying manuscripts using digital photographs can mitigate these issues. Researchers can extract meaningful information from digital representations of the manuscripts without requiring direct physical access by relying on image-based material of the writing surfaces. This method not only drastically cuts the time and expenses connected with the classification process but also dramatically decreases the risks associated with handling delicate artifacts.

One of the critical steps in image-based material analysis is classifying the writing surfaces based on the primary materials, for example, papyrus and parchment. This material classification is difficult due to large inter-class and intra-class variations within materials [23]. Framing this problem in the context of ancient historical manuscripts provides a more significant challenge, primarily due to the degree of degradation of the documents. Previous material classification work has focused on inter-material and texture classification techniques using data sets from non-context 'clean' images [29] and data sets from 'wild', context-set real-world images [4]. Other work has incorporated material, texture, and pattern recognition techniques in specific real-world intra-material classification [6, 25, 46]. There has, however, been little usage of surface material classification techniques set in the context of photographic images of ancient manuscripts. This study uses images from the Dead Sea Scrolls (DSS) collection as a data set to investigate a classification method for materials of the writing surfaces (see Fig. 1).

1.1 Dead Sea Scrolls Collection

The ancient historical manuscripts in the DSS collection originate between the third century BCE and the first century CE and were primarily written on parchment (animal skin) and papyrus (made from the pith of the papyrus plant), with copper serving as the only exception [37]. Recent research on the DSS has focused on author identification, scroll dating, and handwriting analysis [9–11, 32]. Material scientists use spectroscopy, scanning electron microscopy, microchemical testing, and micro and macro x-ray fluorescence imaging to analyze the materials of these manuscripts [27, 33, 45]. These studies answer queries about the DSS's provenance and archaeometry and are predicated on the accurate identification of the underlying material. Due to factors like cost, personnel availability, potential manuscript damage, lack of technology, and time, using such techniques might not always be practical. The digitized images can be helpful in this situation. The underlying periodic and regular patterns in the material persist

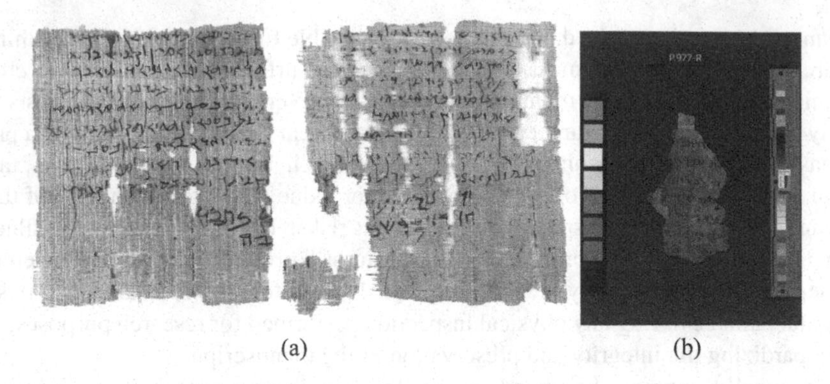

Fig. 1. (a) Color image of plate X102 of the Dead Sea Scrolls collection containing three papyrus fragments. Distinctive striations can be seen in both the vertical and horizontal orientations. Damage is evident on the edges and within each fragment. (b) A multi-spectral color image of a single parchment fragment from plate 977. Color calibration panels, scale bars, and plate labels are visible in the image. Original image source: IAA [37]. These two images are adapted from the earlier work of the same authors [36].

despite deterioration and damage to the manuscripts over time, obstructing accurate and reasonable classification and may serve as the foundation of a classification system ((see Fig. 1)). Testing the precision of such a system can assist in determining whether conventional material analysis methods used on the DSS and other ancient manuscripts have the potential to be complemented by such a system.

1.2 Pattern Recognition Techniques

In the case of texture classification, a dedicated shape feature may prove to be more appropriate and convenient, particularly when considering the limited size of the available training data set. The initial research presented here employs a method in which the regular underlying periodic patterns inherent in the writing surface are used to classify the material of the manuscripts. Different feature vectors are constructed to capture these patterns [36]. This study also presents a hierarchical k-means clustering framework that extends the work using color and texture features. Finally, a deep neural network is shown as an exploratory investigation.

In the initial work, the feature vectors are compared to determine which can categorize the writing materials of the manuscript fragments with the most significant degree of accuracy. The discrete 2-dimensional Fourier transform (2DFT) is the foundation upon which these feature vectors are constructed. From the perspective of computer vision and texture analysis, feature vectors, including using the 2DFT, have produced results that stand alone and complement spatially focused methods [21]. For example, the 2DFT distinguishes between material textures and objects in non-contextual [2,5] and contextual images [7] in conjunction with a standalone classifier [22] or input to a neural network [16,26]. Additionally, adding the 2DFT to Compositional Pattern Producing Networks (CPPNs) has improved the accuracy of texture reproduction because it can match and capture high-frequency details. [42]. Furthermore, the Fourier transform offers a more straightforward method for modeling textures than Markov Random

Field modeling [20]. To classify the materials on which ancient historical manuscripts like the DSS are written, 2DFT feature vectors provide an appealing way to distinguish between the innate textures and patterns in the writing surface.

As an extension to the initial works [36], the current study also investigates the feasibility of a pattern recognition system in a hierarchical k-means clustering framework. The corresponding objective is to group images of the fragments highly likely to originate from the same manuscript in the same cluster. These methods require no ground truths and fewer data samples than deep learning, which is more feasible for analyzing historical documents. Previous research in content-based image retrieval has shown that combining different image properties such as color, texture, and shape improved retrieval results compared to using isolated properties [14]. Color, texture, and shape properties can be extracted from images and clustered using a k-means clustering algorithm [39]. Rasheed et al. [34,35] attempted to automate fragment reassembling from fractured ancient potteries by classifying the fragments into similar groups based on color and texture properties. They achieved high success rates in both studies, where fragments were grouped based on their color properties, and these groups were then further divided into other sub-groups based on their texture properties. Based on these findings, this study will examine the possibility of finding similar patterns and characteristics in color and texture properties among DSS fragments originating from the same physical source. Images of the fragments are utilized as a dataset with Fig. 2 showing an example image. A hierarchical k-means clustering framework is proposed to explore similarities in color and texture properties among fragments.

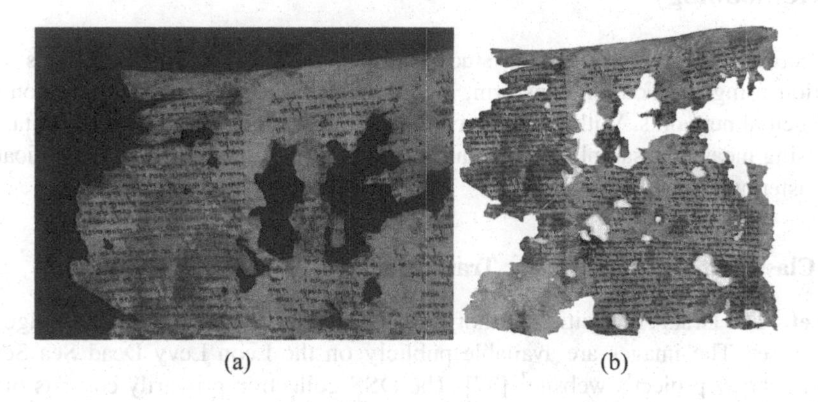

Fig. 2. (a) A multi-spectral color image with partial view of fragment 1 from plate 155-1. A clear indication of erosion can be observed from damages, uneven colors, and text degradation visible in the material. Material loss is also visible from the holes within the fragment. (b) RGB-color image of the full plate 155-1. Comparing this full plate image to its multi-spectral fragment image on the left, the texture is less detailed in the full plate due to lower resolution and image quality.

Finally, this study investigates deep learning-based models as an exploratory and comparative analysis. Previous work in assembling fragments from the DSS proposed a deep neural network to match papyrus fragments originating from the same physical source by utilizing their thread patterns [1]. This network achieves high accuracy by using large papyrus scrolls. To reach a similar goal, a self-supervised Siamese model can also be effective without using knowledge transfer from a substantial dataset [31]. Their Siamese model obtains a *top-1* accuracy of 0.73. Although the deep learning approach can be intriguing, it entails some limitations. Deep learning requires a large amount of labeled data to train a model [48]. Most scrolls are written on parchments without enough ground truths on the material sources. The current study uses the VGG16 model [38], similar to previous work [31], in a self-supervised Siamese setup. In the self-supervised configuration, the *pretext-task* trains the model to generalize to knowledge for the *goal-task*. The model will learn which parchment patches belong to the same fragment. This deep learning-based exploration is confined to parchment patches, assuming that the total dataset size is insufficient to generalize the model to a level that can match parchment and papyrus fragments with minimal samples.

This article makes three main contributions that open several doors for further in-depth analysis of the materials of DSS using pattern recognition techniques:

- A novel feature extraction technique using the two-dimensional Fourier transform to identify repetitive patterns within the manuscript surfaces of the DSS collection to classify a limited number of images.
- A hierarchical k-means clustering framework to explore similarities in color and texture properties among a large number of fragments of the DSS collection.
- A convolutional neural network model, in a self-supervised Siamese setup, uses most of the DSS collection's parchment fragments to classify the material sources.

2 Methodology

This section will briefly present this study's three pattern recognition techniques: Classification using the Fourier transform, hierarchical k-means clustering, and convolutional neural networks. Within these techniques, the section will discuss the data, preprocessing measures, sampling techniques, feature vector construction, Classification, and clustering steps.

2.1 Classification Using Fourier Transform

Dataset. The Israel Antiquities Authority (IAA) kindly provided the DSS images for the data set. The images are available publicly on the Leon Levy Dead Sea Scrolls Digital Library project's website[1] [37]. The DSS collection primarily consists of two materials: mostly parchment with a small percentage written on papyrus and one copper plate as an exception. This study uses a binary classification task on parchment and papyrus. Most scrolls have undergone some degradation from aging or improper handling. A scroll frequently has missing pieces, with only a few disconnected fragments left for analysis. The remaining pieces range in size and condition; some are quite small. A variety of spectral bands are available for some of the images. The initial study uses two sets of scroll images: Set-1 are regular color photographs, and Set-2 are composite images of individual spectral band images.

[1] https://www.deadseascrolls.org.il/explore-the-archive.

Figures 1a and 1b show examples from Set-1 and Set-2 respectively. The sets are used independently. The two sets of images differ from one another. Set-1's images are all 96 dpi, but due to the photography techniques and the curators' fragment arrangement, there are differences in each image's size and distance from the subject. This makes it possible to evaluate the suggested methods using imperfect data. Set-2 images, however, are consistent: 7216 × 5412 pixels in size, 1215 dpi, and with a bit depth of 24. The measurement of the camera's distance to the subject is constant throughout the entire Set-2 of images. Some Set-2 images contain more than one fragment of scroll due to the nature of the arrangement by the curators. Some Set-2 images are only partial images of a larger fragment. The images used for this first experiment are a subset of the complete collection. The subset was made based on two criteria: first, to represent as many of the various textures present throughout the collection as possible, and second, to increase the amount of material from which samples could be taken. The images comprise 23 parchment fragments and 10 papyrus fragments, totaling 33 fragments in each set. Both sets of images utilized the same fragments (For additional details, please see Table 7 in Appendix A).

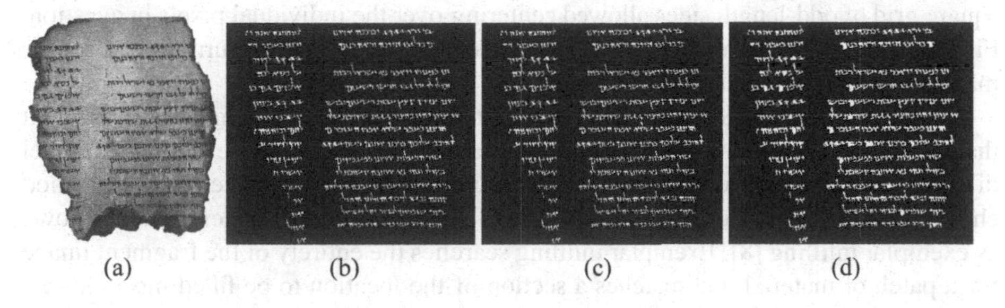

Fig. 3. (a) Color image of plate 976. (b) The binarized (color-inverted) image shows the ink (text) from plate 976. The binarization is obtained using the BiNet [11]. (c) A binary text mask was taken from plate 976 in un-dilated form. (d) A binary text mask was taken from plate 976 after applying dilation to capture all written markings.

Image Preprocessing. The first step in preprocessing is to identify each image fragment. This step is challenging, especially for the Set-2 images that have measurement markings and color reference panels (Fig. 1b). The fragments in an image are identified from the background, other fragments, and the reference markings using automated k-means clustering and by hand. Then, each fragment is extracted for individual processing.

Clean images of the writing surface material within each fragment are required to extract features. Any regular and periodic patterns within the text are prevented from influencing patterns found in the material by removing the text. The text and gaps created by damage within the boundaries of each fragment are filled to provide more sample material, considering the demand for clean background material and the limited availability of DSS material. Previous work using deep learning [11] has produced binary masks of only the visible text in each image (Fig. 3b). As each binary text image overlays onto the original image, these masks identify text locations within each

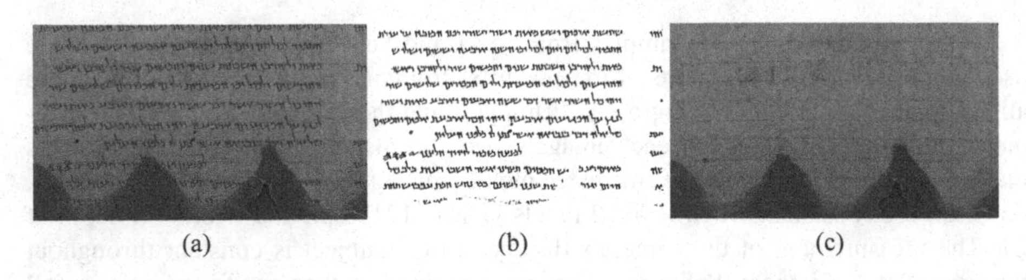

Fig. 4. (a) A zoomed view *before* the filling process; taken from the multi-spectral image of plate 974 (Set-2). (b) A binarized image showing the ink (text) from plate 974. The binarization is obtained using the BiNet [11]. (c) A zoomed view *after* the filling process; taken from the multi-spectral image of plate 974 (Set-2).

fragment. These masks are made robust by dilation; one pixel of text in the mask is expanded to a 3×3 pixel square (For further details, please see Figs. 3c–d). This is a necessary alteration as the masks were used to analyze the written text of the scrolls, having been designed to capture no background material strictly. The 3×3 expansion was chosen for three reasons. Firstly, to increase the coverage of the mask. Secondly, a square grid of odd-length sides allowed centering over the individual pixels in question. Finally, 3×3 is the minimum of such an expansion to leave more surrounding source material available for analysis.

As a result, the written text's outline might still be present in some cases. Any text that was missed by the original binary masks can be captured with the help of text pixel dilation. The primary goal is to preserve the material's regular patterns. The method chosen for filling these locations was selected to keep the surface patterns and is known as exemplar infilling [8]. Exemplar infilling searches the entirety of the fragment image for a patch of material that matches a section of the location to be filled most closely, based on the sum of squared differences. This patch then fills that section (Fig. 4). Each location is filled using different, closely matched patches of specified size (9×9 pixels). A 9×9 patch size was selected as a balance between the computational time needed to search for and fill using such a patch size (a function of the types of images used in this study), the relatively thin text mask areas for which the patch would fill, and the flexibility the exemplar infilling could provide when filling the text space. Future research will be needed to optimize this parameter. A new patch is searched for after every iteration, as the closest matched patches may change once the filling process starts. Additionally, the edges of each scroll fragment are where most of the damage is located. Samples are taken from the fragment's interior to lessen the degradation's impact on the classification step. This is accomplished by only considering the area of the fragment's largest inscribed rectangle (Fig. 5a). Within this region, samples are collected for the construction of feature vectors.

Sampling. In a 5×5 grid covering the sample area, 25 samples of size 256×256 pixels are taken at regular intervals. This size is required by the specifications for using the 2DFT and the images' sizes. In order to provide majority voting evidence and relatively thorough coverage of the surface material of each fragment while balancing the resources required to generate these samples and the image sizes, 25 samples are used. For additional research on this topic, variations on these options are advised. The distance between samples will vary depending on the size of each sample area. As a result, for smaller sample areas, the samples overlap, and for larger sample areas, there is unused space between the samples. The inconsistent fragment size distribution makes this situation unavoidable.

Feature Vectors. Each sample image's saturation values are taken from an image's RGB to HSV conversion. Each pixel is given a value between 0 and 1 during extraction to indicate its saturation level. When using the 2DFT for image classification, it has been demonstrated that the HSV and saturation values enhance performance [25,46]. The 2DFT is then applied. The DC component is centralized, and the log transform of the absolute values is taken (log2DFT) to display the spectrum visually (Fig. 5b). The log2DFT representation is then partitioned into $n \times n$ non-overlapping sections (Fig. 5c). This study uses $n = 7$. As there has been little previous work concerning the use of the log2DFT on ancient historical manuscripts, this value was a subjective decision made to balance the feature vector detail level with the feature vector length and acts as a parameter to be optimized in future work.

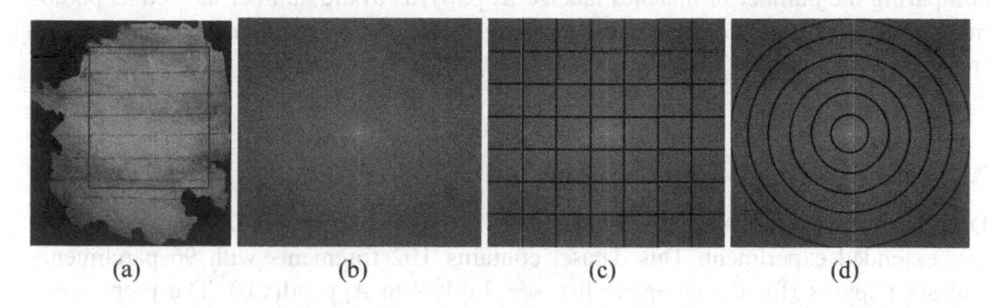

Fig. 5. (a) Sample area from a fragment extracted from the color image of plate 1039-2 (Set-1). The largest internal inscribed rectangle (by area) is found. Samples for feature vector creation will be taken from this area at evenly spaced intervals. (b) Visual representation of the log2DFT applied to a parchment-image sample. (c) The grid creates the primary feature vectors. The mean and standard deviation of the pixel values in each grid area are used as the basis for the feature vectors. (d) The concentric ring division is used in the construction of secondary feature vectors. The mean and standard deviation of the pixel values in the ring area are used as the basis for the feature vectors. Images are taken from [36].

The mean values in each section are calculated, and the values are concatenated to produce a feature vector for the sample image. This process is repeated for the standard deviation of the pixels, resulting in two separate feature vectors for a sample (mean feature vector as MFV and standard deviation feature vector as SDFV). These are known as the primary feature vectors. In addition, three secondary feature vectors are proposed, making five in total. The first is a feature vector based on dividing the log2DFT into six concentric rings (Fig. 5d). The mean and standard deviation of the pixel values in each ring are calculated and separately concatenated to create two feature vectors, just like the primary feature vectors. The rotational variance of the log2DFT is used to determine whether to include a concentric ring feature vector, with the orientation of the sample impacting classification performance. An ultimate feature vector developed by [7] is tested. This method uses each pixel's magnitude and phase angle from a log2DFT. The magnitude acts as a weighted vote and deposits the pixel into one of nineteen phase angle bins, evenly spaced from 0 to 2π. The value across all bins is normalized to one and concatenated to produce a 1×19 feature vector for the sample.

Classification. The same method is used to handle independently each of the five proposed feature vectors. Each sample's associated feature vector and ground truth, parchment or papyrus, are kept in a dictionary, and only one feature vector is considered. There are 825 distinct entries in the dictionary. Then, a leave-one-out strategy is used. A fragment's entire 25 feature vectors are purged from the dictionary. These feature vectors are essentially invisible at this time. The remaining 25 vectors stored in the dictionary are compared to each of the 25 removed vectors to determine the closest match. The Euclidean distance between the two vectors is the foundation for this match. The associated ground truth is noted once the closest feature vector in the dictionary has been identified. A percentage of belief regarding the fragment's material is provided by comparing the number of matches labeled as papyrus to the number labeled as parchment. The fragment is categorized according to the more significant percentage. All fragments are repeated in this manner. At both the fragment and sample levels, the F-scores are computed.

2.2 Hierarchical K-Means Clustering

Data and Preprocessing. A larger dataset is created from the DSS collection for this extended experiment. This dataset contains 102 fragments with 96 parchments and six papyrus (for the complete list, see Table 9 in Appendix B). The preprocessing has been performed in a similar way explained for the initial experiment in Subsect. 2.1. Larger fragments are prioritized to maximize the patch extraction area; fragments, where approximately up to four non-overlapping patches can be extracted, are included in this study. This process is illustrated in Fig. 6.

Feature Extraction. Color properties are considered by assuming that fragments originating from the same physical source share similar color properties. Texture properties are considered for two reasons: First, to distinguish between primary surface materials, e.g., between parchment and papyrus, and second, to distinguish within parchment or papyrus. Both color and texture features are extracted from all patches. In this study, color features are defined from color moments, and texture features are defined as a combination of a co-occurrence matrix of the local binary pattern and Gabor wavelet transforms.

Color Moments. Color moments are statistical measurements that represent the dominant features of the color distribution of an image [41]. Various image retrieval studies have utilized color moments in image retrieval tasks and showed high-performance measurements when grouping images based on these color features [30,44]. This study uses the following color moments:

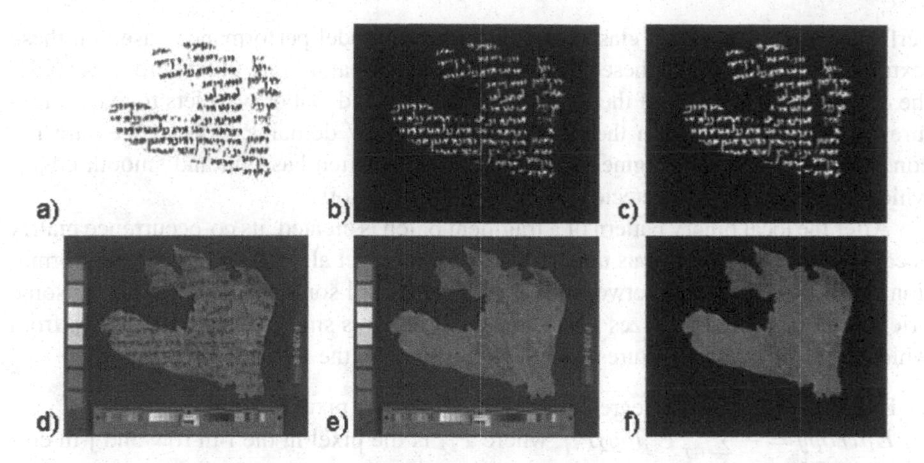

Fig. 6. Exampler infilling preprocessing stages visualized for plate 228-1, fragment-1. The binarized image from BiNet [12] (a) is inverted (b) and dilated with a disk size of 3 × 3 pixels to create a mask (c). The mask removes the text from the multi-spectral image (d) utilizing the exampler infilling method (e). Then, the background information is removed from the image to capture the fragment(f) solely.

1. Mean: indicates the average color of an image:
   $E_i = \frac{1}{N}\sum_{N}^{j=1} p_{ij}$, where N is the number of pixels, i is the i-th color channel, j is the j-th pixel in the image and p_{ij} is the j-th pixel in the i-th color channel.
2. Standard deviation (sd): takes the square root of the variance in the color distribution and describes how much the pixel values variate between each other:
   $\sigma_i = \sqrt{(\frac{1}{N}\sum_{N}^{j=1}(p_{ij} - E_i)^2)}$, where E_i is the mean pixel value of the i-th color channel.
3. Skewness: provides information about the shape of the color distribution by evaluating how asymmetric the distribution is:
   $s_i = \sqrt[3]{(\frac{1}{N}\sum_{N}^{j=1}(p_{ij} - E_i)^3)}$
4. Kurtosis: measures how extreme the tail of the distribution is:
   $k_i = \sqrt[4]{(\frac{1}{N}\sum_{N}^{j=1}(p_{ij} - E_i)^4)}$

Texture Information. A texture classification study suggested that combining different texture features can improve the accuracy of texture classification [3]. Zhu et al. [47] introduced a method where they fused texture features extracted from the spatial domain using a grey-level co-occurrence matrix (GLCM) and from the frequency domain using Gabor wavelets to classify cashmere and wool fibers. Similar findings were further supported by Tou et al. [43] as their model performance also improved when combining features obtained from GLCM and Gabor wavelets for classifying different textural materials. Another study tested texture classification on stone texture by combining local binary patterns, GLCM, and edge detection features [15]. The local binary pattern of an image was created from which its co-occurrence matrix was calculated, followed by the Sobel edges. Combining these features showed improved model performance over different classifiers compared to model performance based on these texture features. Based on these findings, two texture features are utilized for this study: the co-occurrence matrix of the local binary pattern and Gabor wavelets to extract texture information from both the spatial and frequency domains. Sobel edges are not considered because the fragments' surface material often has thin and smooth edges, which are not accurately detected by Sobel edge detectors.

After the local binary pattern of a fragment patch is created, its co-occurrence matrix is calculated. This matrix was introduced by Haralick et al. [19] and provides information about the correlation between pixel values up until some neighbor pixels for some orientation. This study utilizes seven statistical features similar to the study [15], from which five are Haralick features, and the other two are the mean and variance.

1. Entropy: evaluates the degree of randomness of the pixel distribution:
   $Entropy = -\sum_{i,j} P_{ij} log_2 P_{ij}$, where P_{ij} is the pixel in the i-th row and j-th column of the matrix.
2. Energy: takes the square root of the angular second-moment, which is a measurement for the homogeneity of an image. Fewer dominant pixel transitions are more homogeneous:
   $Energy = \sum_{i,j}(P_{ij})^2$.
3. Contrast: measures how much local variation is present in the image. The contrast is lower for images with similar pixel intensity values:
   $Contrast = \sum_{i,j} |i - j|^2 P_{ij}$.
4. Correlation: measures the number of linear dependencies in the image:
   $correlation = \sum_{i,j} \frac{(i_{\mu_i})(j-\mu_i)P_{ij}}{\sigma_i \sigma_j}$.
5. Homogeneity: measures how similar the pixel intensity values are:
   $homogeneity = \sum_{i,j} \frac{P_{ij}}{1+|i-j|}$.
6. Mean: calculates the mean value of the matrix:
   $Mean = \sum_{i,j} i P_{ij}$
7. Variance: measures how different the data values are from the mean:
   $variance = \sum_{i,j} P_{ij}(i - \mu)^2$

Gabor wavelets are a set of different Gabor filters modulated by a Gaussian envelope [18]. The patches have been sliced from a 256 × 256 pixel dimension to an 86 × 86 pixel dimension to capture the periodic patterns. For this study, the frequencies 0.05 and 0.25 from the x- and y-coordinates one and three are evaluated as these showed satisfactory results by empirical testing. Like the Haralick features, the Gabor wavelets are calculated in eight directions to assure rotation-invariant features. The coordinates are computed for positions one to three. The patch is first converted to a grey image and then convolved with the kernel to obtain the frequency-time domain of the patch. From this domain, the mean and variance are calculated. The Haralick and Gabor features are concatenated in a one-dimensional vector, resulting in 98 texture features for each patch.

Feature Normalization and Space Reduction. To scale the properties accordingly, a min-max normalization scaler is applied, which takes care of scaling each feature within a specific range according to $X_{scaled} = X_{std} * (X_{max} - X_{min}) + X_{min}$, where X_{std} is the standard deviation of the feature calculated from the feature set and X_{min}

and X_{max} are the minimum and maximum values of the desired range. In this study, a range of $[0, 1]$. Following the data normalization, a principal component analysis (PCA) is performed (variance ratio is set to 95% threshold) on both the color and texture feature sets to retain only the features that affect the results.

Clustering. A hierarchical k-means framework is implemented to group fragments with similar color and texture properties. The clustering algorithm is known for partitioning an unlabelled dataset into k different clusters [28]. The objective is to minimize the total intra-cluster variance using some distance metric. This study utilizes the Euclidean distance metric, which makes the objective function $\sum_{j=1}^{k}\sum_{i=1}^{n}||x_i - c_j||^2$, where k is the number of clusters, n is the number of data points, x_i is data point i and c_j is the centroid of cluster j. Figure 7 shows a schematic of the proposed framework. An optimal k number is selected using ten-fold cross-validation for color and texture properties. Table 10 in Appendix B shows the cross-validation results of color cluster one. The main framework is implemented as follows:

1. The patches are partitioned according to their color properties in k_1 color clusters.
2. One fragment has eight patches; not all eight patches will be clustered in the same cluster. A majority vote has been performed to determine to which cluster the fragment belongs.
3. For each fragment, all eight patches are saved in the cluster the fragment belongs to. These patches are used as data points for texture clustering.
4. From the resulting k_1 color clusters, each cluster is forwarded to the same k-means algorithm separately using the texture features of the patches. Each color cluster is divided into a different amount of texture clusters. Doing this for all k_1 color clusters will result in k_2 texture clusters.
5. Again, a majority vote is performed to determine which cluster the fragment belongs to. These resulting texture clusters are used for evaluation.

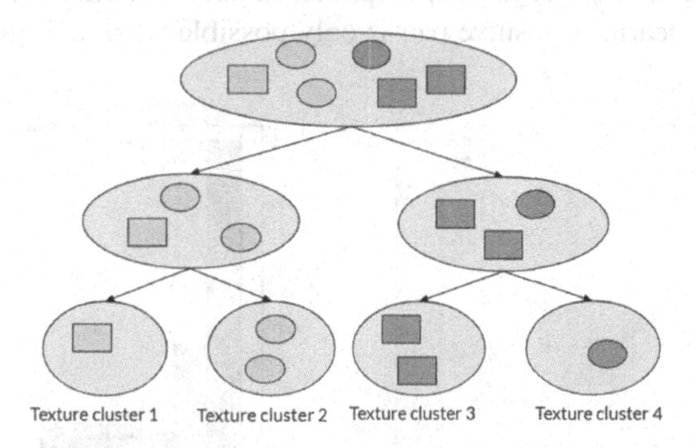

Fig. 7. Schematic graph of the hierarchical k-means framework. The objects represent the fragments, and the shapes illustrate different texture properties.

2.3 Convolutional Neural Networks

Data and Preprocessing. For this final experiment, more than 15000 fragment images were initially chosen from around 800 parchment plates of the DSS collection, each with a dpi of 1215. To isolate the parchment images, the initial classification work (see Subsect. 2.1) is utilized [36]. From the parchment images, fragment masks are created for each image by converting color images to binary images, and then after noise removal by finding the object (connected component) near the center of the image (see Fig. 8a–c). The VGG16 [31] was designed for images of dimension 224 × 224 pixels. The sizes of the fragment images are substantially larger. Besides, for the *pretext-task*, the model should learn whether a pair of patches belongs to a specific fragment. Given both facts, this experiment extracts patches of dimension 256 × 256 pixels from the fragment images. This process can be seen as overlaying the image of the fragment with a grid in Fig. 8d. Each block in that grid has the size of the patch dimension. Moreover, each block represents a unique patch.

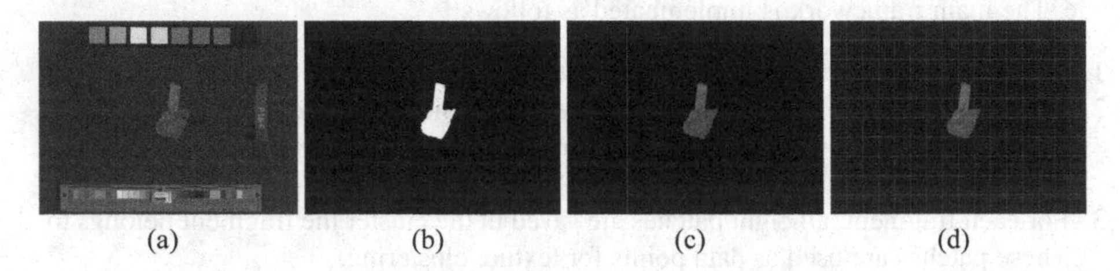

Fig. 8. (a) Fragment 1 from plate 18. (b) Final Mask (the white pixel blob with its centroid closest to the center of the image). (c) Isolated fragment. (d) Grid applied.

To ensure that the dataset only contains patches with fragment information, completely black patches are removed from the dataset. However, there are still patches with a majority of black pixels. Therefore, a minimum amount of non-black pixels should be decided so that the model still has access to sufficient fragment information per patch and that the size of the dataset stays appropriate. In addition, the Siamese model needs pairs of patches to learn. A positive pair is only possible when a fragment has at least

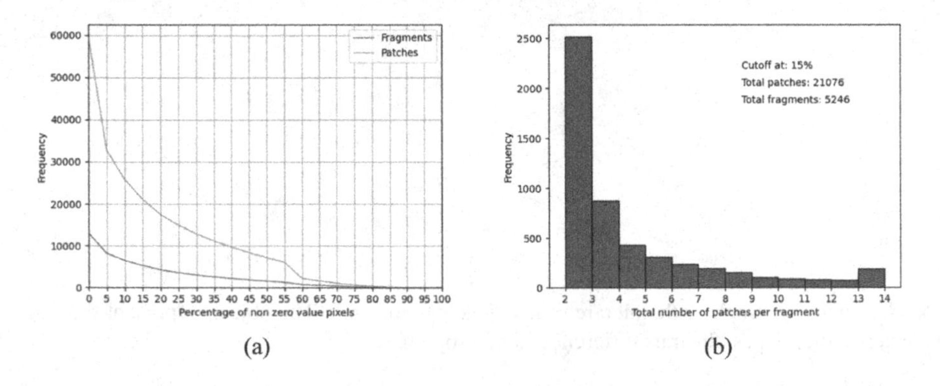

Fig. 9. (a) Frequency of patches & fragments per criteria. (b) Patches at cutoff 15%.

two patches that fulfill the minimum non-black pixels. Figure 9a shows the histogram of the percentage of non-zero pixels for patches and fragments. After visually inspecting Fig. 9a, a minimal non-black pixels percentage of 15 is decided for the dataset. With this percentage, the dataset size becomes 21076 patches. Figure 9b illustrates the distribution of patches per fragment for the given cutoff percentage.

Model. In a Siamese setup, two equal feature extractors are used. The output(s) of these two feature extractors are then combined. This experiment uses VGG16 as the feature extractor. Leaky Relu replaces the activation layers of VGG16. Moreover, the final dense layers are removed and replaced by a flattening layer to obtain a 1D representation. This 1D representation allows applying an Euclidian distance layer, calculating the difference between feature extractors and their output. The combined output goes through two final dense layers of size 512 each and ends at a sigmoid function. The value 0 of the sigmoid function represents a dissimilar pair, and the value 1 illustrates a similar pair. The model learns what pair is similar (1) and dissimilar (0). Figure 10 shows a schematic of the model. The model is trained for 100 epochs with Binary Cross Entropy Loss and Adam optimizer. The model trains with a static learning rate of 0.00001 and batch size 16 based on previous works [24]. The model performance is calculated using accuracy, precision, recall, and the F-score.

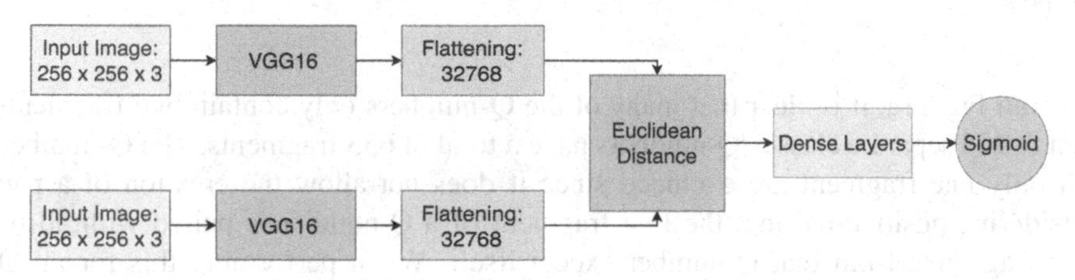

Fig. 10. Diagram of the Siamese VGG16 [31].

Self-supervised Learning. Since the provided DSS dataset does not have a ground truth of which fragments belong to the same source, this experiment refers to self-supervised learning. During self-supervised learning, the model learns its *goal-task* through its *pretext-task*.

Pretext-Task. To ensure that the model is not affected by an unbalanced dataset during training, the positive and negative pairs are created for the *pretext-task*. Since the model trains on a binary classification task, an optimal distribution contains 50% positive pairs and 50% negative pairs. For this experiment, a positive pair is set as a pair of patches that originate from the same fragment. Utilizing the bins shown in Fig. 9b, the model goes through each odd fragment in all the bins. A fragment is represented by the set of patches it consists of. The model pairs each instance of that set with another instance from the same set until all possible unique pairs are obtained. Similarly, negative pairs are also created with different sets. This dataset finally contains 27842 positive pairs and 27842 negative pairs. The pairs are split into 80% training/validation data and 20% test data.

Goal-Task. As stipulated previously, the goal task exists out of the model recognizing which pair of fragments originate from the same source. However, no ground truth exists to assess the model's performance on the *goal-task*. However, Q-numbers are available and can be assumed as individual sources for the DSS. Figure 11a illustrates how often a Q-number contains two or more fragments.

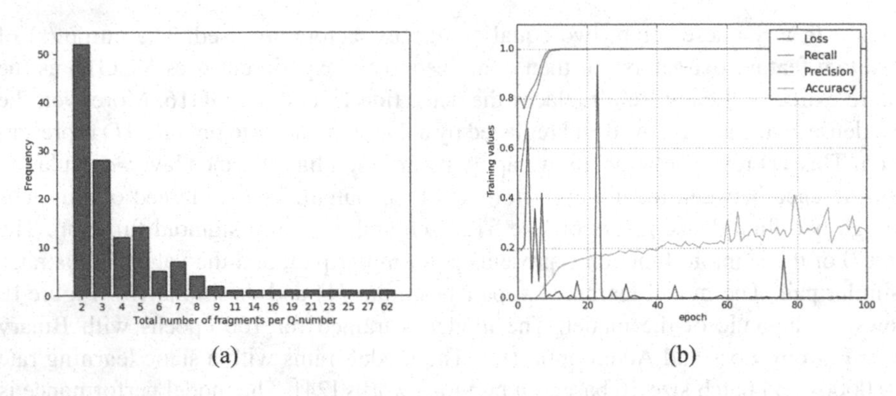

(a) (b)

Fig. 11. (a) Number of fragments for a Q-number. (b) Training performance of the model over 100 epochs.

From Fig. 11a, it is clear that many of the Q-numbers only contain two fragments. In total, 133 separate classes/Q-numbers have a total of 653 fragments. The Q-numbers with only one fragment are excluded since it does not allow the creation of a pair. Considering positive pairing, the first fragment of a Q-number is paired with all the other fragments from that Q-number except itself. When performing this for all Q-numbers, 520 positive pairs are obtained. To acquire an equal probability of confronting the model with either a positive or negative pair, 520 negative pairs are also obtained. The negative pairs are found by pairing fragments that do not belong to the same Q-number until reaching the size of the positive pairs. Again, each pair is unique. These steps result in a balanced dataset containing 1040 pairs of fragments for testing the *goal-task*.

3 Results

3.1 Classification Using Fourier Transform

For the first experiment, 33 fragments (23 parchments, 10 papyrus) were used. The overall classification success percentage for the image types using the primary feature vectors MFV and SDFV is 90.9% for color images and 97.0% for multi-spectral images, respectively. Table 1 shows the confusion matrices for the results of the primary feature vectors. Tables 2 and 3 show the precision, recall, and F-score for classification at the fragment and sample levels for the primary feature vectors. The three secondary feature vectors showed less successful results (For further details, please see Table 8 in Appendix A).

Table 1. Confusion matrix (%) for the MFV and SDFV. Adapted from [36].

Image type	True class	MFV		SDFV	
		Classified as		Classified as	
		Parchment	Papyrus	Parchment	Papyrus
Color	Parchment	**100.0**	0.0	**95.7**	4.3
	Papyrus	30.0	**70.0**	20.0	**80.0**
Multi-spectral	Parchment	**100.0**	0.0	**100.0**	0.0
	Papyrus	10.0	**90.0**	10.0	**90.0**

3.2 Hierarchical K-Means Clustering

In this study, 102 fragments were evaluated with 53 Q-numbers, of which 19 have multiple fragments and 34 have single fragments. The 19 Q-numbers with multiple fragments were evaluated to determine the accuracy of the color clusters. The ones with single fragments were not included since it is not concrete with which fragments they share similar color properties and therefore cannot be compared. The results are shown in Table 4. The texture clustering results were evaluated and summarized in Table 5. All 53 Q-numbers are taken into consideration for evaluation; however, only the 19 Q-numbers assigned to multiple fragments were considered for calculating the accuracy. Classification between parchment and papyrus was also taken into consideration. Results showed that parchment and papyrus fragments were clustered separately, achieving a performance of 100% for material classification. Finally, the average accuracy scores of the color and texture clustering are 77% and 68%, respectively. The material classification accuracy is 100%.

Table 2. Precision, Recall and F-score for classification at the fragment level [36].

Image	Mean Feature Vector (MFV)			
	Material	Precision	Recall	F-score
Color	Parchment	0.88	1	0.94
	Papyrus	1	0.70	0.82
Multispc.	Parchment	0.96	1	**0.98**
	Papyrus	1	0.90	0.95
Image	Std. Dev. Feature Vector (SDFV)			
Color	Parchment	0.92	0.96	0.94
	Papyrus	0.89	0.80	0.84
Multispc.	Parchment	0.96	1	**0.98**
	Papyrus	1	0.90	0.95

Table 3. Precision, Recall and F-score for classification at the sample level [36].

Image	Mean Feature Vector (MFV)			
	Material	Precision	Recall	F-score
Color	Parchment	0.85	0.91	0.88
	Papyrus	0.77	0.64	0.70
Multispc.	Parchment	**0.93**	**0.96**	**0.95**
	Papyrus	0.91	0.84	0.87
Image	Std. Dev. Feature Vector (SDFV)			
Color	Parchment	0.85	0.89	0.87
	Papyrus	0.72	0.64	0.68
Multispc.	Parchment	**0.91**	**0.96**	**0.93**
	Papyrus	0.89	0.78	0.83

3.3 Convolutional Neural Networks

During this experiment, multiple configurations of the Siamese VGG16 were tested. The best-performing model is presented here. After training the model for 100 epochs on the *pretext-task*, the performance on the training set is shown in Fig. 11b. The test performance of the model on the *pretext-task* and the *Goal-task* is illustrated in Table 6. Table 6 indicates that the model performs well on the *pretext-task* when considering all performance scores. Both recall (0.84) and precision (0.86) scores are similar. Therefore, the effect of false positives and false negatives on the model performance is equivalent.

Moreover, the accuracy shows that the model was correct in identifying each pair 85% of the time. Lastly, the F-score demonstrates minimal negative effects of incorrectly identified pairs. For the *Goal-task*, the results show that the recall score of 0.45 deviates substantially from the precision score of 0.75. Hence, the effect of false negatives is higher than false positives. The model more often falsely predicted a pair as negative than positive. Furthermore, the F-score of 0.56 indicates that the incorrectly identified pairs negatively affect the model. However, the accuracy of 0.66 shows that the model's predictions aligned with the actual Q-numbers 66% of the time.

Table 4. Accuracy scores for the 19 Q-numbers clustered over all four color clusters.

Q-number	total fragments	cluster 1	cluster 2	cluster 3	cluster 4	Average accuracy
4Q186	3	1	**2**			0.67
4Q424	2	**1**		1		0.50
4Q59	2	**2**				1.00
4Q177	3	**2**	1			0.67
4Q176	2	**1**		1		0.50
4Q405	4	**4**				1.00
11Q10	4	**4**				1.00
4Q13	5	1		**4**		0.80
4Q32	3		**2**	1		0.67
4Q385; 4Q285a	5	**2**	1	2		0.40
4Q419;4Q420	3		1	**2**		0.67
11Q5	12	3		**5**	4	0.42
Sdeir4	2	2				1.00
4Q427	2			**2**		1.00
4Q212	4			**4**		1.00
4Q214a; 4Q214b	3		1	1	1	0.33
11Q1	5			**5**		1.00
8Hev1	3				**3**	1.00
5/6 Hev7	3				**3**	1.00

Table 5. Results of texture clustering for all four color clusters.

Color cluster	1	2	3	4
Number of texture clusters	7	6	4	3
Q-numbers $\geq$ two fragments	7	2	7	3
Number of fragments	19	4	24	10
Average accuracy score	0.60	0.75	0.75	0.60

Table 6. Model performance on the test data.

	recall	precision	F-score	accuracy
pretext task	0.84	0.86	0.85	0.85
goal task	0.45	0.75	0.56	0.66

4 Discussions

To classify material used in the DSS, this study first presented a binary 2DFT-based method as pilot work. This method demonstrated a relatively high level of performance with regard to the overall classification percentage ($\approx$97% compared to $\approx$91% successful classification for MS and color images, respectively) across both primary feature vectors (based on Fourier-space grid representation) used in conjunction with the multi-spectral images. The results from the mean feature vector were also marginally better than those from the standard deviation feature vector. This was especially true for classification accuracy determined by sample-level F-score values. Papyrus images had lower recall and F-scores in their classification than parchment samples, making them more prone to misclassification than parchment images (see Tables 2 and 3).

The proposed 2DFT-based method was most accurate when combined with multi-spectral images and the MFV. The same applies to color images. This outcome might be helpful for upcoming work on manuscripts that have yet to be photographed to use multi-spectral equipment. The performance difference between the multi-spectral and the color images can be attributed to several factors. Comparing multi-spectral images to color images, for instance, reveals more discernible patterns, periodic frequencies, magnitudes, and details in the materials of the image. This results from the various light wavelengths emphasizing various material details. For the log2DFT method, recombining images at multiple wavelengths may capture more features that distinguish each material.

The multi-spectral images also had higher resolution, which helped the visual-based technique distinguish between materials more clearly. The spectral band that produces the best results can be used in additional research by looking at each spectral band image separately. This strategy may have an impact on choices made about photographing other manuscripts. Regarding the F-score, the MFV performed at least as well as the SDFV. This may imply that measures of the spread between pixel values in grid sections have a lower ability to discriminate between different objects than measures that consider the value of the pixels.

Future research may examine which frequencies and related magnitudes enhance material discrimination. For the first test in this study, a dictionary of feature vectors was compiled using various images. These photos were selected because, in the authors' opinion, they accurately captured the range of textures found in papyrus and parchment and adequately represented the entire DSS collection. Papyrus samples made up a smaller percentage of the collection than parchment ones. The papyrus samples consequently proved more challenging to classify accurately. The collection can be better represented by expanding the dictionary set, producing better classification outcomes from a more extensive and well-balanced data set, and significantly improving the classification of papyrus fragments. As a result, more examples would be found and possible closer matches to novel samples. The measured voting process would require more votes to confirm a classification, affecting the belief percentage per sample.

Binary text masks helped fill the images. However, it might only sometimes be possible to access these materials. Some binary images failed to capture all of the text on the manuscripts in a small minority of instances, most notably affecting some papyrus manuscripts. The classification of the papyrus fragments, which showed more residue text post-fill, performed marginally worse than expected. The study employed a fixed log2DFT feature vector grid and a fixed sample size. Instead, using wedges radiating from the origin of a centered 2DFT in the feature construction process is a possible pathway for improving this method, mainly if texture patterns exhibit edges or lines in a particular direction, such as with papyrus [40]. Concerning the modified weighted bin feature vector, which showed success in the study by [7], this was particularly evident, achieving a low 10% success rate for papyrus images (For further details, see the tables in Appendix A).

The initial study used 33 images. To extend this work, 53 different Q-numbers with 102 fragments were utilized to evaluate the performance of the hierarchical k-means clustering framework. The fragments were first clustered based on their color properties in four different sets, and the Q-number evaluation showed a 77% accuracy score. Afterward, the fragments were clustered based on their texture properties in 20 different texture sets, resulting in a 68% accuracy score. Performance on material classification is also measured, and the framework succeeded in clustering parchment and papyrus fragments separately, resulting in a 100% accuracy score. These results are promising and suggest that a hierarchical k-means clustering framework has the potential to cluster fragments of the DSS utilizing their color and texture properties. However, these results are based on evaluating the Q-numbers assigned to the fragments by domain experts and palaeographers. These assignments may not always suggest that all fragments from the same manuscript (Q-number) originated from the same physical source. The possibility that fragments with the same Q-number do not originate from the same physical source should be kept open.

The study finally introduces a neural network-based model where more than 15000 fragment images were used from the DSS collection. When referring to the test data on the *pretext-task*, all the performance measures showed that the model performs well. The effect of false predictions is minimal regarding a recall score of 0.84 and a precision score of 0.86. The F-score of 0.85 summarizes this finding. Interestingly, the model obtained an accuracy of 0.86. It was capable of correctly identifying the 11,136 pairs 86% of the time. The test on the *goal-task* shows a substantial difference between the recall score of 0.45 and the precision score of 0.75. This difference indicates that the model more often falsely predicted a pair of fragments as not belonging to the same Q-number than belonging to the same Q-number. The F-score summarizes the negative effect of falsely identified pairs in terms of a score of 0.56. However, the model's accuracy is 66%. Therefore, the model's prediction aligns with the assigned Q-numbers 66% of the time for the 1040 fragment pairs. This difference lies mainly in the model predicting fragments not originating from the same source while they are according to the Q-numbers. Again, the Q-numbers are not ground truth for physical material sources. Due to the lack of proper ground truth about the materials, this study utilized the Q-numbers as pseudo-ground truths for material classification. This scenario is not perfect, but it opens doors for future digital paleography experiments.

5 Conclusions

This study used three different techniques in classifying the DSS images. The first one is a 2DFT-based feature vector technique in the binary classification of the surface material of the DSS, achieving a performance of up to $\approx$97%. It provides a precise and accessible method of categorizing the content of old historical manuscripts without additional labeled data (as in the case of neural networks) or destructive techniques (for instance, some chemical analyses). The first experiment offers a first attempt at performing material classification of such manuscripts using a 2DFT technique. The initial manuscript investigation may go faster if the writing surface material can be quickly categorized. The straightforward method described here might serve as a starting point for settling any argument regarding the composition of a DSS fragment. It could be used with older historical manuscripts as well. In the second one, by building upon and developing the first system, this study extends the classification work by proposing a hierarchical k-means clustering framework using the color and textures of the fragments. Finally, in the third technique, the study utilizes a convolutional neural network model in a self-supervised Siamese setup and uses patches of the DSS fragment images. All three methods demonstrate the potential to help answer more specialized questions. Examples may include intra-material classification to provide evidence for writer identification, localization, differing production techniques, and manuscript dating. The consequences of gaining such insight by substituting the proposed plans are threefold: preserving delicate ancient manuscripts from further degradation, relatively low-cost implementable methods, and additional extendable tools in gathering evidence to help conclude the questions surrounding the production of such manuscripts.

Acknowledgements. The authors would like to thank Mladen Popović, PI of the European Research Council (EU Horizon 2020) project: The Hands that Wrote the Bible: Digital Palaeography and Scribal Culture of the Dead Sea Scrolls (HandsandBible 640497), who allowed work with the data and provided valuable inputs and the labels for the materials. Finally, we are grateful to the Israel Antiquities Authority (IAA) for the high-resolution images of the Dead Sea Scrolls, courtesy of the Leon Levy DSS Digital Library; photographer: Shai Halevi.

Appendix A

Table 7. Plate Numbers of the Fragments used in the study. Adapted from [36].

Plate number	No. of Fragments	Material
1039-2	5	Parchment
155-1	1	Parchment
193-1	1	Parchment
228-1	1	Parchment
269	1	Parchment
489	1	Parchment
641	1	Parchment
974	1	Parchment
975	1	Parchment
976	4	Parchment
977	4	Parchment
978	1	Parchment
979	1	Parchment
5-6Hev45	1	Papyrus
641	1	Papyrus
X100	1	Papyrus
X106	1	Papyrus
X130	1	Papyrus
X207	3	Papyrus
X304	1	Papyrus
Yadin50	1	Papyrus

Table 8. Confusion matrix (%) for Mean Concentric Ring Feature Vector (MCRFV), Standard Deviation Concentric Ring Feature Vector (SDCRFV), and Weighted Bin Feature Vector (WBFV). Adapted from [36].

Image type	True class	MCRFV		SDCRFV		WBFV	
		Classified as		Classified as		Classified as	
		Parchment	Papyrus	Parchment	Papyrus	Parchment	Papyrus
Color	Parchment	**100.0**	0.0	**100.0**	0.0	**100.0**	0.0
	Papyrus	40.0	**60.0**	20.0	**80.0**	**100.0**	**0.0**
Multi-spectral	Parchment	**87.0**	13.0	**91.0**	8.7	**95.7**	4.3
	Papyrus	70.0	**30.0**	50.0	**50.0**	90.0	**10.0**

Appendix B

Table 9. Table showing all fragments included in this study with their corresponding plate- as well as Q-numbers and material classification.

Plate	Fragment	Q-number	Material
5-6Hev45	1	5/5Hev45	papyrus
63-1	1,2,3	8Hev1	parchment
64	1	Mur88	parchment
106	2	4Q267	parchment
109	2,3	4Q186	parchment
113	14	4Q186	parchment
116-1	1,2	4Q427	parchment
122A	1	4Q375	parchment
123	1,2	4Q424	parchment
133	1	4Q197	parchment
140	1	4Q258	parchment
155-1	1	4Q403	parchment
156A-1	1	4Q434	parchment
181	2	4Q416	parchment
193-1	1	4Q542	parchment
200	3	4Q204	parchment
201	2	4Q19;4Q20;4Q21	parchment
215	1	4Q2	parchment
226-1	1	4Q325	parchment
227	1,2	4Q212	parchment
228-1	1,2	4Q212	parchment
232	1	4Q72	parchment
233	2,3,4	4Q32	parchment
256-1	1	4Q28	parchment
262	1,15	4Q59	parchment
263	20	4Q87	parchment
264	2	4Q35; 4Q369; 4Q464	parchment
267	2	4Q385a; 4Q385b; 4Q385c	parchment
269	1	4Q386	parchment
270	2,3,5,7	4Q385; 4Q385a	parchment
273	1	4Q3	parchment
277	1,3,8	4Q177	parchment

(*continued*)

Table 9. (*continued*)

Plate	Fragment	Q-number	Material
278	1	4Q552	parchment
279	1	4Q501	parchment
283	1	4Q380	parchment
284	1	4Q492	parchment
285	1,2	4Q176	parchment
286	1	4Q174	parchment
291	1	4Q164	parchment
300	5	4Q219	parchment
304	4	4Q98g; 4Q238; 4Q281; 4Q468c; 4Q238; 4Q468b	parchment
370	2,3,5	4Q214a; 4Q214b	parchment
488-1	1	4Q414	parchment
489-1	1	4Q418	parchment
497	3,4,8,9	4Q405	parchment
509	1,10,12	4Q419; 4Q420	parchment
607-1	1	11Q14	parchment
623626630	1,2,3,4	11Q10	parchment
641	1	Mur114	papyrus
659-1	1,2,3,4,5	4Q13	parchment
974	1	11Q5	parchment
975	1	11Q5	parchment
976	1	11Q5	parchment
977	1,2,3,4	11Q5	parchment
978	1	11Q5	parchment
979	1	11Q5	parchment
984	1,3	Sdeir 4	parchment
1039-2	1,2,3,4	11Q1	parchment
x207	1,2,3	5/6Hev 7	papyrus
x304	1	5/6Hev27	papyrus

Table 10. Cross validation results of color cluster one. The results show that for seven clusters, the average efficiency increases again after it has been decreasing from cluster one to six. Therefore, seven is set to be the optimal number of texture clusters for color cluster one.

	2 CLUSTERS	3 CLUSTERS	4 CLUSTERS	5 CLUSTERS	6 CLUSTERS	7 CLUSTERS	8 CLUSTERS	9 CLUSTERS	10 CLUSTER
RUN 1	83.33	74.24	69.70	75.76	15.15	**75.76**	53.03	56.06	45.45
RUN 2	87.88	75.76	74.24	65.15	15.15	**71.21**	48.48	48.48	34.85
RUN 3	78.79	84.85	71.21	80.30	69.70	**36.36**	54.55	57.58	46.97
RUN 4	83.33	87.88	74.24	74.24	72.73	**71.21**	62.12	16.67	60.61
RUN 5	80.30	75.76	71.21	28.79	13.64	**68.18**	57.58	62.12	68.18
RUN 6	84.85	74.24	75.76	75.76	62.12	**69.70**	66.67	53.03	53.03
RUN 7	83.33	75.76	75.76	71.21	68.18	**69.70**	57.58	53.03	56.06
RUN 8	87.50	84.85	71.21	65.15	80.30	**63.64**	57.58	60.61	69.70
RUN 9	83.33	84.85	53.03	59.09	59.09	**69.70**	59.09	62.12	65.15
RUN 10	80.30	78.79	77.27	71.21	69.70	**63.64**	74.24	65.15	53.03
AVERAGE	83.29	79.70	71.36	66.67	52.58	**65.91**	59.09	53.49	55.30

References

1. Abitbol, R., Shimshoni, I., Ben-Dov, J.: Machine learning based assembly of fragments of ancient papyrus. J. Comput. Cult. Heritage (JOCCH) **14**(3), 1–21 (2021)
2. Bajcsy, R.: Computer description of textured surfaces. IJCAI 572–579 (1973)
3. Barley, A., Town, C.: Combinations of feature descriptors for texture image classification. J. Data Anal. Inf. Process. **02**, 67–76 (2014). https://doi.org/10.4236/jdaip.2014.23009
4. Bell, S., Upchurch, P., Snavely, N., Bala, K.: Material recognition in the wild with the Materials in Context Database. In: Proceedings of the IEEE Computer Society Conference on Computer Vision and Pattern Recognition, 07–12-June, pp. 3479–3487 (2015). https://doi.org/10.1109/CVPR.2015.7298970
5. Bharati, M.H., Liu, J.J., MacGregor, J.F.: Image texture analysis: methods and comparisons. Chemom. Intell. Lab. Syst. **72**(1), 57–71 (2004). https://doi.org/10.1016/j.chemolab.2004.02.005
6. Camargo, A., Smith, J.S.: Image pattern classification for the identification of disease causing agents in plants. Comput. Electron. Agric. **66**(2), 121–125 (2009). https://doi.org/10.1016/j.compag.2009.01.003
7. Cevikalp, H., Kurt, Z.: The Fourier transform based descriptor for visual object classification. Anadolu Univ. J. Sci. Technol. Appl. Sci. Eng. **18**(1), 247 (2017). https://doi.org/10.18038/aubtda.300419
8. Criminisi, A., Pérez, P., Toyama, K.: Region filling and object removal by exemplar-based image inpainting. IEEE Trans. Image Process. **13**(9), 1200–1212 (2004). https://doi.org/10.1109/TIP.2004.833105
9. Dhali, M.A., He, S., Popović, M., Tigchelaar, E., Schomaker, L.: A digital palaeographic approach towards writer identification in the dead sea scrolls. In: Proceedings of the 6th International Conference on Pattern Recognition Applications and Methods 2017-Janua(Icpram), ICPRAM 2017, pp. 693–702 (2017). https://doi.org/10.5220/0006249706930702
10. Dhali, M.A., Jansen, C.N., de Wit, J.W., Schomaker, L.: Feature-extraction methods for historical manuscript dating based on writing style development. Pattern Recogn. Lett. **131**, 413–420 (2020). https://doi.org/10.1016/j.patrec.2020.01.027
11. Dhali, M.A., de Wit, J.W., Schomaker, L.: BiNet: degraded-manuscript binarization in diverse document textures and layouts using deep encoder-decoder networks. arXiv (2019). http://arxiv.org/abs/1911.07930

12. Dhali, M.A., de Wit, J.W., Schomaker, L.: Binet: degraded-manuscript binarization in diverse document textures and layouts using deep encoder-decoder networks. arXiv preprint arXiv:1911.07930 (2019)
13. Drira, F.: Towards restoring historic documents degraded over time. In: Second International Conference on Document Image Analysis for Libraries (DIAL 2006), pp. 8-pp. IEEE (2006)
14. Duan, G., Yang, J., Yang, Y.: Content-based image retrieval research. Physics Procedia **22**, 471–477 (2011). https://doi.org/10.1016/j.phpro.2011.11.073. 2011 International Conference on Physics Science and Technology (ICPST 2011)
15. Ershad, S.: Texture classification approach based on combination of edge & co-occurrence and local binary pattern. arXiv (2012)
16. Franzen, F., Yuan, C.: Visualizing image classification in Fourier domain. In: Proceedings, 27th European Symposium on Artificial Neural Networks, Computational Intelligence and Machine Learning, ESANN 2019, vol. 27, pp. 535–540 (2019)
17. Freedman, J., van Dorp, L., Brace, S.: Destructive sampling natural science collections: an overview for museum professionals and researchers. J. Nat. Sci. Collections **5**, 21–34 (2018)
18. Gabor, D.: Theory of communication. J. Inst. Electr. Eng. **93**(3), 429–457 (1946)
19. Haralick, R., Shanmugam, K., Dinstein, I.: Textural features for image classification. IEEE Trans. Syst. Man Cybern. **3**(6), 610–621 (1973)
20. Hassner, M., Sklansky, J.: The use of Markov Random Fields as models of texture. Comput. Graph. Image Process. **12**(4), 357–370 (1980). https://doi.org/10.1016/0146-664X(80)90019-2
21. Hu, X., Ensor, A.: Fourier spectrum image texture analysis. In: International Conference Image and Vision Computing New Zealand 2018-Novem(1), pp. 1–6 (2019). https://doi.org/10.1109/IVCNZ.2018.8634740
22. Hui, S., Zak, S.H.: Discrete Fourier transform based pattern classifiers. Bull. Pol. Acad. Sci. Tech. Sci. **62**(1), 15–22 (2014). https://doi.org/10.2478/bpasts-2014-0002
23. Kalliatakis, G., et al.: Evaluating deep convolutional neural networks for material classification. arXiv 2 (2017)
24. Kandel, I., Castelli, M.: The effect of batch size on the generalizability of the convolutional neural networks on a histopathology dataset. ICT Express **6**(4), 312–315 (2020)
25. Kliangsuwan, T., Heednacram, A.: FFT features and hierarchical classification algorithms for cloud images. Eng. Appl. Artif. Intell. **76**, 40–54 (2018). https://doi.org/10.1016/j.engappai.2018.08.008
26. Kumar, Y., Jajoo, G., Yadav, S.K.: 2D-FFT based modulation classification using deep convolution neural network. In: 2020 IEEE 17th India Council International Conference (INDICON), pp. 1–6. IEEE (2020)
27. Loll, C., Quandt, A., Mass, J., Kupiec, T., Pollak, R., Shugar, A.: Museum of the Bible Dead Sea Scroll Collection Scientific Research and Analysis. Final Report, Art Fraud Insights online (2019)
28. MacQueen, J.: Some methods for classification and analysis of multivariate observations. In: Proceedings of the Fifth Berkeley Symposium on Mathematical Statistics and Probability, vol. 1, pp. 281–297 (1967)
29. Matsuyama, T., Miura, S.I., Nagao, M.: Structural analysis of natural textures by Fourier transformation. Comput. Vis. Graph. Image Process. **24**(3), 347–362 (1983). https://doi.org/10.1016/0734-189X(83)90060-9
30. Mufarroha, F., Anamisa, D., Hapsani, A.: Content based image retrieval using two color feature extraction. In: Journal of Physics: Conference Series, vol. 1569, p. 032072 (2020). https://doi.org/10.1088/1742-6596/1569/3/032072
31. Pirrone, A., Beurton-Aimar, M., Journet, N.: Self-supervised deep metric learning for ancient papyrus fragments retrieval. Int. J. Document Anal. Recogn. (IJDAR) 1–16 (2021)

32. Popović, M., Dhali, M.A., Schomaker, L.: Artificial intelligence based writer identification generates new evidence for the unknown scribes of the dead sea scrolls exemplified by the great Isaiah scroll (1qisaa). PLoS ONE **16**(4), e0249769 (2021)
33. Rabin, I.: Archaeometry of the dead sea scrolls. Dead Sea Discoveries **20**(1), 124–142 (2013). http://www.jstor.org/stable/24272914
34. Rasheed, N., Md Nordin, J.: Archaeological fragments classification based on RGB color and texture features. J. Theor. Appl. Inf. Technol. **76**(3), 358–365 (2015). https://doi.org/10.1016/j.jksuci.2018.09.019
35. Rasheed, N., Md Nordin, J.: Classification and reconstruction algorithms for the archaeological fragments. J. King Saud Univ. Comput. Inf. Sci. **32**(8), 883–894 (2020). https://doi.org/10.1016/j.jksuci.2018.09.019
36. Reynolds, T., Dhali, M.A., Schomaker, L.: Image-based material analysis of ancient historical documents. In: Proceedings of the 12th International Conference on Pattern Recognition Applications and Methods, ICPRAM, pp. 697–706. INSTICC, SciTePress (2023). https://doi.org/10.5220/0011743700003411
37. Shor, P.: The Leon Levy Dead Sea scrolls digital library. The digitization project of the dead sea scrolls. Scholarly Commun. **2**(2), 11–20 (2014). https://doi.org/10.1163/9789004264434_003
38. Simonyan, K., Zisserman, A.: Very deep convolutional networks for large-scale image recognition. arXiv preprint arXiv:1409.1556 (2014)
39. Sleit, A., Abu Dalhoum, A., Qatawneh, M., Al-Sharief, M., Al-Jabaly, R., Karajeh, O.: Image clustering using color, texture and shape features. KSII Trans. Int. Inf. Syst. **5**, 211–227 (2011). https://doi.org/10.3837/tiis.2011.01.012
40. Sonka, M., Hlavac, V., Boyle, R.: Image Processing, Analysis, and Machine Vision. Thomson-Engineering, 4th edn. (2015)
41. Stricker, M., Orengo, M.: Similarity of color images. In: Niblack, W., Jain, R.C. (eds.) Storage and Retrieval for Image and Video Databases III, vol. 2420, pp. 381–392. International Society for Optics and Photonics, SPIE (1995). https://doi.org/10.1117/12.205308
42. Tesfaldet, M., Snelgrove, X., Vazquez, D.: Fourier-CPPNs for image synthesis. In: Proceedings of the 2019 International Conference on Computer Vision Workshop, ICCVW 2019, pp. 3173–3176 (2019). https://doi.org/10.1109/ICCVW.2019.00392
43. Tou, J.Y., Tay, Y.H., Lau, P.Y.: Gabor filters and grey-level co-occurrence matrices in texture classification. In: MMU International Symposium on Information and Communications Technologies (2007)
44. Varish, N., Singh, P.: Image retrieval scheme using efficient fusion of color and shape moments. In: Patgiri, R., Bandyopadhyay, S., Balas, V.E. (eds.) Proceedings of International Conference on Big Data, Machine Learning and Applications. LNNS, vol. 180, pp. 193–206. Springer, Cham (2021). https://doi.org/10.1007/978-981-33-4788-5_16
45. Wolff, T., et al.: Provenance studies on Dead Sea scrolls parchment by means of quantitative micro-XRF. Anal. Bioanal. Chem. **402**, 1493–1503 (2012). https://doi.org/10.1007/s00216-011-5270-2. www.axp.tu-berlin.de
46. Wu, X., et al.: Fourier transform based features for clean and polluted water image classification. In: Proceedings of the International Conference on Pattern Recognition 2018-Augus, pp. 1707–1712 (2018). https://doi.org/10.1109/ICPR.2018.8546306
47. Zhu, Y., Huang, J., Wu, T., Ren, X.: Identification method of cashmere and wool based on texture features of GLCM and Gabor. J. Eng. Fibers Fabrics **16**, 1–7 (2021). https://doi.org/10.1177/1558925021989179
48. Zohuri, B., Moghaddam, M.: Deep learning limitations and flaws. Mod. Approaches Mater. Sci. **2** (2020). https://doi.org/10.32474/MAMS.2020.02.000138

Author Index

M. De Marsico et al. (Eds.): ICPRAM 2023, LNCS 14547, p. 151, 2024.
https://doi.org/10.1007/978-3-031-54726-3

Printed in the United States
by Baker & Taylor Publisher Services

Printed in the United States
by Baker & Taylor Publisher Services